SpringerBriefs on Cyber Security Systems and Networks

Editor-in-Chief

Yang Xiang, Digital Research & Innovation Capability Platform, Swinburne University of Technology, Hawthorn, VIC, Australia

Series Editors

Liqun Chen, Department of Computer Science, University of Surrey, Guildford, Surrey, UK

Kim-Kwang Raymond Choo, Department of Information Systems, University of Texas at San Antonio, San Antonio, TX, USA

Sherman S. M. Chow, Department of Information Engineering, The Chinese University of Hong Kong, Hong Kong, Hong Kong

Robert H. Deng, School of Information Systems, Singapore Management University, Singapore, Singapore

Dieter Gollmann, E-15, TU Hamburg-Harburg, Hamburg, Hamburg, Germany

Kuan-Ching Li, Department of Computer Science & Information Engineering, Providence University, Taichung, Taiwan

Javier Lopez, Computer Science Department, University of Malaga, Malaga, Spain

Kui Ren, University at Buffalo, Buffalo, NY, USA

Jianying Zhou, Infocomm Security Department, Institute for Infocomm Research, Singapore, Singapore

The series aims to develop and disseminate an understanding of innovations, paradigms, techniques, and technologies in the contexts of cyber security systems and networks related research and studies.

It publishes thorough and cohesive overviews of state-of-the-art topics in cyber security, as well as sophisticated techniques, original research presentations and in-depth case studies in cyber systems and networks. The series also provides a single point of coverage of advanced and timely emerging topics as well as a forum for core concepts that may not have reached a level of maturity to warrant a comprehensive textbook.

It addresses security, privacy, availability, and dependability issues for cyber systems and networks, and welcomes emerging technologies, such as artificial intelligence, cloud computing, cyber physical systems, and big data analytics related to cyber security research. The mainly focuses on the following research topics:

Fundamentals and theories

- Cryptography for cyber security
- Theories of cyber security
- Provable security

Cyber Systems and Networks

- Cyber systems security
- Network security
- Security services
- Social networks security and privacy
- Cyber attacks and defense
- Data-driven cyber security
- Trusted computing and systems

Applications and others

- Hardware and device security
- Cyber application security
- Human and social aspects of cyber security

More information about this series at http://www.springer.com/series/15797

Kwangjo Kim · Harry Chandra Tanuwidjaja

Privacy-Preserving Deep Learning

A Comprehensive Survey

 Springer

Kwangjo Kim 🔟
School of Computing
Korea Advanced Institute of Science
and Technology (KAIST)
Daejeon, Korea (Republic of)

Harry Chandra Tanuwidjaja 🔟
School of Computing
Korea Advanced Institute of Science
and Technology (KAIST)
Daejeon, Korea (Republic of)

ISSN 2522-5561   ISSN 2522-557X (electronic)
SpringerBriefs on Cyber Security Systems and Networks
ISBN 978-981-16-3763-6   ISBN 978-981-16-3764-3   (eBook)
https://doi.org/10.1007/978-981-16-3764-3

This Springer imprint is published by the registered company Springer Nature Singapore Pte Ltd.
The registered company address is: 152 Beach Road, #21-01/04 Gateway East, Singapore 189721, Singapore

To our friends and families for their lovely support.

Preface

This monograph aims to give a survey on the state of the art of Privacy-Preserving Deep Learning (PPDL), which is considered to be one of the emerging technologies by combining classical privacy-preserving and cryptographic protocols with deep learning in a systematic way.

Google and Microsoft announced a big investment in PPDL in the early 2019, followed by the announcement of "Private Join and Compute", an open-source PPDL tools that is based on Secure Multi-Party Computation (Secure MPC) and Homomorphic Encryption (HE) on June 2019 by Google. One of the main issues in PPDL is about its applicability, e.g., to understand the gap between the theory and practice exists. In order to solve this, there are many advances relying on the classical privacy-preserving method (HE, secure MPC, differential privacy, secure enclaves, and its hybrid) and together with deep learning. The basic architecture of PPDL is to build a cloud framework that enables collaborative learning while keeping the training data on the client device. After the model is fully trained, the privacy during the sensitive data exchange or storage must be strictly preserved and the overall framework must be feasible for the real applications.

This monograph plans to provide the fundamental understandings for privacy preserving and deep learning, followed by comprehensive overview of the state of the art of PPDL methods, suggesting the pros and cons of each method, and introducing the recent advances of the federated learning and split learning-based PPDL called as Privacy-Preserving Federated Learning (PPFL). In addition, this monograph gives a guideline to general people and students, and practitioners who are interested to know about PPDL and also helping early-stage researcher who wants to explore PPDL area. We hope that the early-stage researchers can grasp the basic theory of PPDL, understand the pros and cons of current PPDL and PPFL methods, addressing the gap between theory and practice in the most recent approach, so that they can propose their own method later.

Daejeon, Korea (Republic of)   Kwangjo Kim
March 2021   Harry Chandra Tanuwidjaja

Acknowledgements

This monograph was partially supported by Institute for Information & communications Technology Promotion (IITP) grant funded by the Korea Government Ministry of Science and ICT (MSIT) (No. 2017-0-00555, Towards Provable-secure Multi-party Authenticated Key Exchange Protocol based on Lattices in a Quantum World and No. 2019-0-01343, Regional Strategic Industry Convergence Security Core Talent Training Business.)

The authors sincerely appreciate the contribution of the lovely alumni and members of Cryptology and Information Security Lab (CAISLAB), Graduate School of Information Security, School of Computing, KAIST, including but not limiting to Rakyong Choi, Jeeun Lee, Seunggeun Baek, Muhamad Erza Aminanto, and Edwin Ayisi Opare for their volunteering help and inspiring discussions.

In particular, we would like to mention our sincere gratitude to Indonesia Endowment Fund for Education, Lembaga Pengelola Dana Pendidikan (LPDP) for supporting Harry Chandra Tanuwidjaja during his Ph.D. study at KAIST.

We gratefully acknowledge the editors of this monograph series on security for their valuable comments and the Springer for giving us the opportunity to write and publish this monograph.

Finally, we also are very grateful for our families for their strong support and endless love *at the foot of the spiritual Kyerong Mt. which is located at the center of the Republic of Korea.*

Contents

Acronyms

ABE	Attribute-based Encryption
ADPPL	Adaptive Differential Privacy-Preserving Learning
AI	Artificial Intelligence
ANN	Artificial Neural Network
CNN	Convolutional Neural Network
DNN	Deep Neural Network
DP	Differential Privacy
EDPR	European General Date Protection Regulation
FE	Functional Encryption
FHE	Fully Homomorphic Encryption
GAN	Generative Adversarial Networks
Gaussian DP	Gaussian Differential Privacy
GC	Garbled Circuit
GCD	Greatest Common Divisor
GRU	Gated Recurrent Unit
HBC	Honest-but-Curious
HE	Homomorphic Encryption
IBE	Identity-based Encryption
IoT	Internet of Things
kNN	K-Nearest Neighbors
LDP	Local Differential Privacy
LSH	Locality-Sensitive Hashing
LSTM	Long Short-Term Memory Networks
LWE	Learning With Errors
MDP	Markov Decision Processes
MLaaS	Machine Learning as a Service
MPC	Multi-party Computation
NN	Neural Network
OT	Oblivious Transfer
PATE	Private Aggregation of Teacher Ensembles
PoC	Privacy of Client
PoM	Privacy of Model

PoR	Privacy of Result
PP	Privacy Preserving
PPDL	Privacy-Preserving Deep Learning
PPFL	Privacy-Preserving Federated Learning
PPRL	Privacy-Preserving Reinforcement Learning
ReLU	Rectified Linear Unit
Ring-LWE	Ring Learning With Errors
RL	Reinforcement Learning
RNN	Recurrent Neural Networks
RPAT	Randomized Privacy-Preserving Adjustment Technology
Secure MPC	Secure Multi-Party Computation
SGD	Stochastic Gradient Decent
SGX	Software Guard Extensions
SIMD	Single Instruction Multiple Data
SVM	Support Vector Machine
SVM-RFE	SVM-Recursive Feature Elimination
TEEs	Trusted Execution Environments

Chapter 1
Introduction

Abstract This chapter provides an introductory discussion why the privacy of each living person is so important and discusses the deployment of privacy-preserving deep learning due to its development of artificial intelligence and the performance advantage of their various implementation.

1.1 Background

Every individuals can have their own private information, e.g., name, date of birth, gender, place of birth, parents' name, hobby, taste, clinical data, nationality, and personal identification number, etc. after their birth. Single private data of a individual is difficult to specify him or her due to lack of private data. Combination of partial private data on a individual can specify a person easily. If a person passed away, we do not need to consider the validity of his/her privacy in general. Information holders can decide their data open to the public, but some sensitive information must be kept or processed in securely and privately not to reveal their privacy which can be interpreted as *leave me alone or keep me in private*.

To get the access grant from the remote server over the Internet, the clients at first have to submit their unique personal identification data to the server at the initial registration, then the server must keep this sensitive data to be safe area in an unrecognizable (encrypted) form assuming that we can trust the server completely or no insider attacks. Next step is the challenge-response or Sigma protocol executed between two-party: a client and the server to get the reliable connection to the server using a unique challenging data. The sensitive personal data concatenated with unique timestamp or random number can be used for the identification process in every access attempt to prevent from the replay attack by eavesdroppers. In this identification we need to remember that the illegal hackers can break into the server intentionally and try to get the private date of all clients in order to collect money illegally or execute the impersonation which behaves like legitimate clients to the server. By this hacking, the privacy issues may arise independent to the serious cautions of the clients.

K. Kim and H. C. Tanuwidjaja, *Privacy-Preserving Deep Learning*,
SpringerBriefs on Cyber Security Systems and Networks,
https://doi.org/10.1007/978-981-16-3764-3_1

1.2 Motivation

Cloud computing environments can be modelled like three-party protocol; data owner, cloud server and statistic analyst. Due to the increase of the personal data day-by-day, data owner wants to save their important data to the cloud severs that are committed to maintain user's data correctly and safely. Statistic analyst wants to get on-demand statistics from all data stored at cloud servers for the sake of public safety, policy, census, etc. Data owners tend to reluctantly submit their private data to a third party like statistic analyst. A risk of data leakage will happen due to compromised cloud computing server-side too. Users choose not to store their confidential data in cloud because they worry about that insiders at cloud server may look at or leak their private data.

In order to convince users for their data security and privacy, an approach to use privacy-preserved data must be considered by sending encrypted private data to the cloud severs. From the encrypted data stored at cloud servers, statistic analyst has difficult to get correct statistics. In this scenario, we need to consider two goals: privacy and public interest together such as *catch two birds with a stone.* in Chinese proverb and prepare for a special tool to process over the encrypted data by using homomorphic encryption. which is a kind of very special encryption allowing one to perform required calculations such as addition or multiplication on encrypted data.

Cryptography is everywhere in year 2000 which means that every individual can use secure communications and understand the importance of cryptography for personal business over the Internet. Security is used for everywhere in year 2010. Privacy becomes everywhere in year 2015.

When the data is exchanged or communicated between different parties then it's compulsory to provide security to that data so that other parties do not know what data is communicated between original parties. Preserving data means hiding output knowledge of data mining by using several methods when this output data is valuable and private. Mainly two techniques are used for this one is Input privacy in which data is manipulated by using different techniques and other one is the output privacy in which data is altered in order to hide the rules (What is Privacy preserving Technique 2021). If the data under consideration is related with privacy of a person, we say this is Privacy-Preserving (PP).

Nowadays, the Internet is inseparable from human lives. The data exchange rate over the Internet has increased incredibly using Facebook, Twitter or cloud servers. This has lead into privacy issues that need to be solved urgently.

According to Big Data Market Size (2021), the global big data market size was valued at USD 25.67 billion in 2015 and is expected to witness significant growth over the forecast period. The elevating number of virtual online offices coupled with increasing popularity of social media producing an enormous amount of data is a major factor driving growth. Increased internet penetration owing to the several advantages including unlimited communication, abundant information and resources, easy sharing, and online services generates huge chunks of data in everyday life, which is also anticipated to propel demand over the coming years.

Recently introducing EDPR (European General Date Protection Regulation) (2016) in EU mandates strict regulations regarding the storage and exchange of personal identifiable data and data related with personal health requiring authentication, authorization and accountability.

AI(Artificial Intelligence) is to make a computer program infers and behaves like a normal human by computer scientist in the beginning. The term of AI has been changed to Machine learning (ML) which belong to a subset of AI. Due to latest advancement of Deep Neural network, the term of Deep Learning (DL) becomes most popular which belongs to a class of machine learning since DL employs consecutive layers of information processing stages in hierarchical manners for pattern classification or representation learning.

According to Deng and Yu (2014), there are three important reasons for the deep learning prominence recently. First, processing abilities (e.g., GPU units) has been increased sharply. Second, computing hardware is getting affordable, and the third is a recent breakthrough in machine learning research. Shallow and deep learners are distinguished by the depth of their credit assignment paths, which are chains of possibly learnable, causal links between actions and effects. Usually, deep learning plays an important role in image classification results. Besides, deep learning is also commonly used for language, graphical modeling, pattern recognition, speech, audio, image, video, natural language, and signal processing. Now *Machine Learning is everywhere including security and privacy*.

Nowadays, due to global COVID-19 pandemic, every person in the world is very concerned to prevent their privacy from contact-tracing necessary for the public healthcare. Keeping privacy from the hackers also becomes of utmost importance in modern life.

In order to address these issues, several privacy-preserving deep learning techniques, including Secure Multi-Party Computation and Homomorphic Encryption in Neural Network have been developed. There are also several methods to modify the Neural Network, so that it can be used in privacy-preserving environment. However, there is a trade-off between privacy and performance among various techniques.

In this book, we survey state-of-the-art of Privacy-Preserving Deep Learning, evaluate all methods from the point of privacy and performance, compare pros and cons of each approach, and address challenges and issues in the field of privacy-preserving by deep learning including federated learning privacy-preserving.

Note that the earlier version of this monograph was published in Tanuwidjaja et al. (2020), but the contents of this monograph was significantly improved by adding more up-to-date publications with our new analysis.

1.3 Outline

The outline of this monograph is as follows:

This chapter provides an introductory discussion why the privacy of each living person is so important and discusses the deployment of privacy-preserving deep learning due to its development of artificial intelligence and the performance advantage of their various implementation.

Chapter 2 provides a fundamental understanding regarding privacy-preserving technologies, followed by deep learning. We introduce the classical privacy-preserving employing group-based anonymity, cryptographic method, differential privacy and enclaves depending on how to implement privacy-preserving mechanism. We also give a brief concept of deep learning techniques including its outline and basic layers, convolutional neural network, generative adversarial network, support vector machine, recurrent neural network, k-means clustering, and reinforcement learning.

In Chapter 3, we survey the latest publications regarding X-based privacy-preserving deep learning methods based on homomorphic encryption, secure multi-party computation, differential privacy, secure enclaves and their hybrid and summarize the key features of all surveyed publications including learning type and dataset.

Chapter 4 discusses the comparison of all of privacy-preserving deep learning methods, highlighting the pros and cons of each method based on privacy parameters, used specific neural network and dataset type from the point of performance. We also provide our analysis about the weakness of each privacy-preserving deep learning method and our feasible solution to address their weakness.

Chapter 5 introduces the emerging application of privacy-preserving federated learning in a coordinated way among multi-party. We suggest the function specific Privacy-Preserving Federated Learning (PPFL) to provide fairness, integrity, correctness, adaptiveness and flexibility. Application specific PPFL includes mobile devices, medical imaging, traffic flow prediction and healthcare, Android malware detection, and edge computing.

Chapter 6 categorizes the types of the adversarial model on privacy-preserving deep learning based on its behavior, define the major security goals of privacy-preserving deep learning for machine learning as a service, discuss the possible attacks on privacy-preserving deep learning for machine learning as a service, and provide detailed explanations about the protection against the attacks.

Finally in back matter, the concluding remarks are made and interesting further challenges are suggested. We hope that this monograph not only provides a better understanding of PPDL and PPFL but also facilitates future research activities and application development.

References

Big data market size, share & trends analysis report by hardare, by service, by end-use, by region & segment forcasts, 2018–2025, Grand View Research (2021). https://www.grandviewresearch.com/industry-analysis/big-data-industry

Deng L, Yu D (2014) Deep learning: methods and applications. Found Trends Signal Process 7(3–4):197–387

GDPR (2016) Intersoft Consulting. https://gdpr-info.eu

Tanuwidjaja HC, Choi R, Baek S, Kim K (2020) Privacy-preserving deep learning on machine learning as a service—a comprehensive survey. IEEE Access, vol 8, pp 167 425–167 447

What is privacy preserving technique, IGI Global (2021). https://www.igi-global.com/dictionary/privacy-preserving-technique-ppt/58814

Chapter 2
Preliminaries

Abstract This chapter provides a fundamental understanding regarding privacy-preserving technologies, followed by deep learning. We introduce the classical privacy-preserving employing group-based anonymity, cryptographic method, differential privacy and enclaves depending on how to implement privacy-preserving mechanism. We also give a brief concept of deep learning techniques including its outline and basic layers, convolutional neural network, generative adversarial network, support vector machine, recurrent neural network, k-means clustering, and reinforcement learning.

We can classify the classical PP method into four categories, as shown in Fig. 2.1 which are employing group-based anonymity, cryptographic method, differential privacy and enclaves depending on how to implement PP mechanism.

2.1 Classical Privacy-Preserving Technologies

2.1.1 Group-Based Anonymity

In 1983, the famous cryptographer (Chaum 1983) has proposed an idea to make a blind signature using common public key cryptosystem, for instance RSA or DSA that provides message anonymity of a user (bank customer) to a signer (bank) without giving the message content. The blind signature can be used to provide *unlinkability*, which prevents the signer from the linking the blinded message it signs to a later un-blinded version. This blind signature is very useful to make e-cash, e-voting and e-auction *etc.*where a secret holder can provide random message to make a secret communication while keeping the privacy of a user.

This application can be extended to provide the anonymity of a data owner stored at the public data base where someone else is extremely difficult to specify the data owner by introducing some special means of making anonymous.

The concept of k-anonymity was first introduced by Samarati and Sweeney (1998) in 1998 to solve the problem: "Given sensitive personal data, produce the modified data which remains useful while the data cannot specify the corresponding person."

© The Author(s), under exclusive license to Springer Nature Singapore Pte Ltd. 2021
K. Kim and H. C. Tanuwidjaja, *Privacy-Preserving Deep Learning*,
SpringerBriefs on Cyber Security Systems and Networks,
https://doi.org/10.1007/978-981-16-3764-3_2

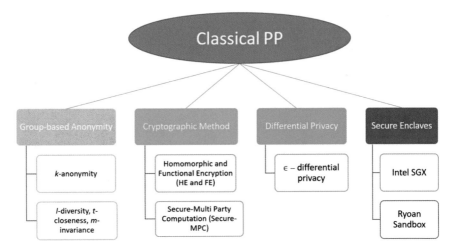

Fig. 2.1 Classical PP classification

Modified data are said to have k-anonymity if the information for any person whose information is in the modified data cannot be distinguished from at least $k-1$ individuals in the modified data. While k-anonymity is a simple and promising approach for group-based anonymization, it is susceptible to attacks such as a homogeneity attack or background knowledge attack (Machanavajjhala et al. 2007) when background knowledge is available to an attacker. To overcome these issues, there are many privacy definitions, such as l-diversity, t-closeness, and m-invariance (Machanavajjhala et al. 2007; Li et al. 2007; Xiao and Tao 2007). The concept of l-diversity means that each equivalent class has at least l distinct values for each sensitive attribute, and t-closeness is a further refinement of l-diversity created by also maintaining the distribution of sensitive attributes. This concepts has limited applications since the data itself is not encrypted, but very simple to implement.

2.1.2   Cryptographic Method

2.1.2.1   Homomorphic and Functional Encryption

While homomorphic encryption, functional encryption, and secure multi-party computation techniques enable computation on encrypted data without revealing the original plaintext, we need to preserve the privacy of sensitive personal data such as medical and health data. One of the earliest milestones to preserving this privacy is to hide these sensitive personal data using data anonymization techniques.

In Rivest et al. (1978) questioned whether any encryption scheme can support computation of encrypted data without knowledge of the encrypted information.

If some encryption scheme supports an arithmetic operation $\circ$ on encrypted data $Enc(m_1 \circ m_2)$, this scheme is called Homomorphic Encryption (HE) on an operation $\circ$. Depending on the computation type of HE supports. it is called partially HE when it supports on the specific computation on encrypted data and Fully HE (FHE) when it supports any kind of computation. For example, the well-knwon RSA encryption holds multiplictive HE. Likewise, a scheme is called as additive HE like Pailler cryptosystem (Pailler 1999) if it supports addition on encrypted data without decryption.

The design of FHE remained as an interesting open problem in cryptography for decades, until Gentry suggested the first innovative FHE in Gentry (2009). Afterwards, there have been a number of studies of HE schemes based on lattices with Learning With Errors (LWE) and Ring Learning With Errors (Ring-LWE) problems and schemes over integers with the approximate Greatest Common Divisor (GCD) problem (Cheon et al. 2013; Van Dijk et al. 2010; Brakerski and Vaikuntanathan 2014, 2011; Gentry et al. 2013; Brakerski et al. 2014; Fan and Vercauteren 2012; Clear and Goldrick 2017; Chillotti et al. 2016, 2020). Earlier works focused on HE were impractical for implementation; however, there are currently open cryptographic algorithm libraries that support HE efficiently, such as HElib, FHEW, and HEEAN (Halevi and Shoup 2014; Ducas and Micciancio 2015; Cheon et al. 2017).

Functional Encryption (FE) was proposed by Sahai and Waters (2005) in 2005 and formalized by Boneh et al. (2011) in 2011. Let a functionality $F : K \times X \rightarrow \{0, 1\}^*$. The functionality F is a deterministic function over (K, X) that outputs $(0, 1)^*$ where K is the key space and the set X is the plaintext space. We say a scheme is FE for a functionality F over (K, X) if it can calculate $F(k, x)$ given a ciphertext of $x \in X$ and a secret key sk_k for $k \in K$.

Predicate encryption (Boneh and Waters 2007) is a subclass of FE scheme with a polynomial-time predicate $P : K \times I \rightarrow \{0, 1\}$ where K is the key space, I is the index set, and the plaintext $x \in X$ is defined as (ind, m); X is the plaintext space, ind is an index, and m is the payload message. As an example, we can define FE functionality $F_{\mathsf{FE}}(k \in K, (\mathsf{ind}, m) \in X) = m$ or $\perp$ depending on whether the predicate $P(k, \mathsf{ind})$ is 1 or 0, respectively. Depending on the choice of the predicate, Identity-based Encryption (IBE) (Shamir 1984; Boneh and Franklin 2001; Waters 2005; Gentry 2006; Gentry et al. 2008; Canetti et al. 2003; Agrawal et al. 2010) and Attribute-based Encryption (ABE) (Sahai and Waters 2005; Goyal et al. 2006) are well-known examples of predicate encryption schemes.

Both FE and HE enable computation over encrypted data. The difference is that the computation output of FE is a plaintext, while the output of HE remains encrypted as HE evaluates the encrypted data without decryption. There is no need for a trusted authority within HE systems. Additionally, HE enables the evaluation of any circuit g over the encrypted data if sk_g is given, but FE enables the computation of only some functions.

2.1.2.2   Secure Multi-party Computation

The purpose of Multi-party Computation (MPC) is to solve the problem of collaborative computing that keeps the privacy of an honest/dishonest user in a group without using any trusted third party. Formally, in MPC, for a given number of participants, $p_1, p_2, \cdots, p_n$, each participant has private data, $d_1, d_2, \cdots, d_n$, respectively. Then, participants want to compute the value of a public function f on those private data, $f(d_1, d_2, \cdots, d_n)$, while keeping their own inputs secret.

The concept of secure computation was formally introduced as secure two-party computation in 1986 by Yao (1986) with the invention of Garbled Circuit (GC). Yao's GC requires only a constant number of communication rounds, and all functions are described as a Boolean circuit. To transfer the information obliviously, Oblivious Transfer (OT) is used. The OT protocol allows a receiver P_R to obliviously select i and receive a message m_i from a set of messages M that belong to a sender party P_S. P_R does not know the other messages in M while P_S does not know the selected message.

Secret sharing is yet another building block for secure MPC protocols, *e.g.*, Goldreich et al. (1987) suggested a simple and interactive secure MPC protocol using the secret-shared values to compute the value. Secret sharing is a cryptographic algorithm where a secret is parted and distributed to each participant. To reconstruct the original value, a minimum number of secret-shared values are required.

Compared with HE and FE schemes, in secure MPC, parties jointly compute a function on their inputs using a protocol instead of a single party. During the process, information about parties' secret must not be leaked. In secure MPC, each party has almost no computational cost with a huge communication cost, whereas the server has a huge computational cost with almost no communication cost in the HE scheme. The parties encrypt their data and send them to the server. The server computes the inner product between the data and the weight value of the first layer and sends the computation result back to the parties. Then, the parties decrypt the results and compute the non-linear transformation. The result is encrypted and transmitted again to the server. This process continues until the last layer has been computed. To apply secure MPC to deep learning, we must handle the communication cost as it requires many rounds of communication between the parties and the server, which is non-negligible.

2.1.3   Differential Privacy

Differential privacy (DP) was first proposed by Dwork et al. (2006) in 2006 as a strong standard to guarantee the privacy of the data. A randomized algorithm A gives ϵ-differential privacy if for all datasets D_1 and D_2 differ in at most one element, and for all subsets $S \in Range(im A)$, where $im A$ denotes the image of A, such that

$$\Pr[\mathscr{A}(D_1) \in S] \le \exp(\epsilon) \cdot \Pr[\mathscr{A}(D_2) \in S].$$

Differential privacy addresses when a trusted data manager wants to release some statistics on the data while the adversary cannot reveal whether some individual's information is used in the computation. Thus, differentially private algorithms probably resist identification and reidentification attacks.

An example of the latest implementation of differential privacy technique was proposed by Qi et al. (2020). They suggested a PP method for a recommender system using Locality-Sensitive Hashing (LSH) technique, which is more likely to assign two neighboring points to the same label. As a result, sensitive data can be converted into less sensitive ones.

2.1.4 Secure Enclaves

Secure enclaves, also known as Trusted Execution Environments (TEEs), are a secure hardware method that provides enclaves to protect code and data from another software on the related platform, including the operating system and hypervisor (Hunt et al. 2018). The concept of an enclave was firstly introduced by Intel (2014), which introduced Software Guard Extensions (SGX), available on Intel processors starting from the Skylake generation (Doweck et al. 2017). Utilizing only SGX for PP is not sufficient from the security and privacy perspectives because the code from the server provider is not trusted. SGX only protects the execution of trusted code on an untrusted platform. A code is called trusted if it is public and users can inspect the code. If the code is private, users cannot be assured that the code does not steal their data. Because of this, SGX needs to be confined to a sandbox to prevent data exfiltration. The most widely used sandbox for SGX is Hunt et al. (2018). The Ryoan sandbox also enables users to verify that the enclave executes standard ML code without seeing the model specifications. As a result, a combination of the SGX and Ryoan sandboxes can guarantee the privacy of both clients and ML models.

2.2 Deep Learning

2.2.1 Outline of Deep Learning

PPDL is a development from the classical DL method. It combines the classical PP method with the emerging DL field. DL itself is a sub-class of machine learning the structure and functionality of that resemble a human brain. The structure of a deep learning model is modelled like a layered architecture. It starts from an input layer and ends with an output layer. Between an input layer and an output layer, there can be one or more hidden layers. The more hidden layers are used, the more accurate the DL model becomes. This is caused by the characteristic of a hidden layer. The output of one hidden layer will become the input of the next hidden layer. If we use more hidden layers, the deeper hidden layer will learn about more specific features. There are several DL methods that are widely used for PP. Based on our research, the most popular DL methods for PP are the Deep Neural Network (DNN), Convolutional Neural Network (CNN), and Generative Adversarial Network (GAN).

2.2.2 Deep Learning Layers

2.2.2.1 Activation Layer

The activation layer, as shown in Fig. 2.2, decides whether the data is activated (value one) or not (value zero). The activation layer is a non-linear function that applies a mathematical process on the output of a convolutional layer. There are several well-known activation functions, such as Rectified Linear Unit (ReLU), sigmoid, and tanh. Because those functions are not linear, the complexity becomes really high if we use the functions to compute the homomorphically encrypted data. Hence, we need to find a replacement function that only contains multiplication and addition operations. The replacement function will be discussed later.

2.2.2.2 Pooling Layer

A pooling layer, as shown in Fig. 2.3, is a sampling layer with the purpose of reducing the size of the data. There are two kinds of pooling: max and average pooling. In HE, we cannot use a max pooling function because we cannot search for the maximum value of encrypted data. As a result, average pooling is the solution to be implemented in HE. Average pooling calculates the sum of values; thus, there is only the addition operation here, which can be used over homomorphically encrypted data.

Fig. 2.2 Activation layer

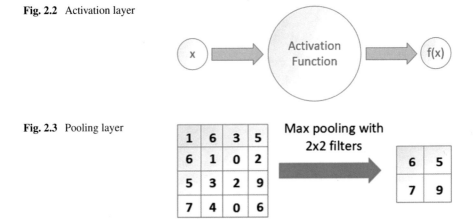

Fig. 2.3 Pooling layer

2.2.2.3 Fully Connected Layer

An illustration of a fully connected layer is shown in Fig. 2.4. Each neuron in this layer is connected to a neuron in the previous layer; thus, it is called a fully connected layer. The connection represents the weight of the feature like a complete binary graph. The operation in this layer is the dot product between the value of the output neuron from the previous layer and the weight of the neuron. This function is similar to a hidden layer in a Neural Network (NN). There is only a dot product function that consists of multiplication and addition function; thus, we can use it over homomorphically encrypted data.

2.2.2.4 Dropout Layer

A dropout layer, shown in Fig. 2.5, is a layer created to solve an over-fitting problem. Sometimes, when we train our machine learning model, the classification result will be too good for some kind of data, showing bias based on the training set. This situation is not ideal, resulting in a large error during the testing period. The dropout layer will drop random data during training and set the data to zero. By doing this iteratively during the training period, we can prevent over-fitting during the training phase.

Fig. 2.4 Fully connected layer

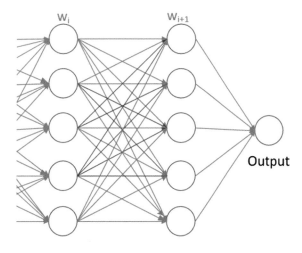

Fig. 2.5 Dropout layer

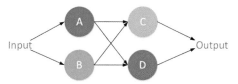

2.2.3 Convolutional Neural Network (CNN)

CNN (LeCun et al. 1999) is a class of DNN usually used for image classification. The characteristic of CNN is a convolutional layer, as shown in Fig. 2.6, the purpose of which is to learn features extracted from the dataset. The convolutional layer has $n \times n$ size, on which we will perform a dot product between neighboring values to make a convolution. As a result, only addition and multiplication occurs in the convolutional layer. We do not need to modify this layer as it can be used for HE data, which are homomorphically encrypted.

2.2.4 Generative Adversarial Network (GAN)

GAN (Goodfellow et al. 2014) is a class of DNN usually used for unsupervised learning. GAN, as shown in Fig. 2.7, consists of two NNs that generate a candidate model and an evaluation model in a zero-sum game framework. The generative model will learn samples from a dataset until it reaches a certain accuracy. On the other hand, the evaluation model discriminates between the true data and the generated candidate model. GAN learns the process by modeling the distribution of individual classes.

2.2.5 Support Vector Machine

A supervised SVM is usually used for classification or regression tasks. If n is the number of input features, the SVM plots each feature value as a coordinate point in n-dimensional space. Subsequently, a classification process is executed by finding the hyperplane that distinguishes two classes. Although SVM can handle a nonlinear decision border of arbitrary complexity, we use a linear SVM since the nature of the dataset can be investigated by linear discriminant classifiers. The decision boundary for linear SVM is a straight line in two-dimensional space. The main computational

Fig. 2.6 Convolutional layer

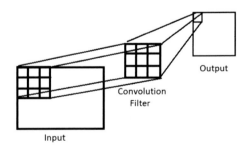

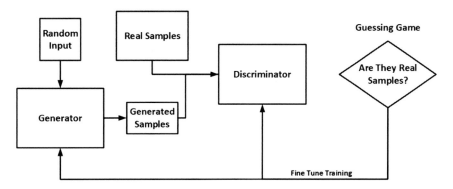

Fig. 2.7 GAN structure

property of SVM is the support vectors which are the data points that lie closest to the decision boundary. The decision function of input vector x as expressed by Eq. (2.1), heavily depends on the support vectors.

$$D(x) = w\mathbf{x} + b \qquad (2.1)$$

$$w = \sum_k \alpha_k y_k x_k \qquad (2.2)$$

$$b = (y_k - wx_k) \qquad (2.3)$$

Equations (2.2) and (2.3) show the corresponding value of w and b, respectively. From Eq. (2.1), we can see that decision function $D(x)$ of input vector $\mathbf{x}$ is defined as the sum of the multiplication of a weight vector and input vector $\mathbf{x}$ and a bias value. A weight vector w is a linear combination of training patterns. The training patterns with non-zero weights are support vectors. The bias value is the average of the marginal support vectors.

SVM-Recursive Feature Elimination (SVM-RFE) is an application of RFE using the magnitude of the weight to perform rank clustering (Guyon et al. 2002). The RFE ranks the feature set and eliminates the low-ranked features which contribute less than the other features for classification task (Zeng et al. 2009).

2.2.6 *Recurrent Neural Network*

Recurrent Neural Networks (RNN) is an extension of neural networks with cyclic links to process sequential information. This cyclic links placed between higher and lower layer neurons which enable RNN to propagate data from previous to current events. This property makes RNN having a memory of time series events

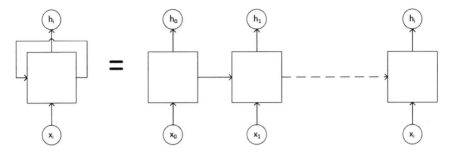

Fig. 2.8 RNN with unrolled topology (Olah 2015)

(Staudemeyer 2015). Figure 2.8 shows a single loop of RNN on the left-side which is comparable to the right-side topology when the loop braked.

One advantage of RNN is the ability connect previous information to present task; however, it cannot reach "far" previous memory. This problem is commonly known as long-term dependencies. Long-Short Term Memory Networks (LSTM) introduced by Hochreiter and Schmidhuber (1997) to overcome this problem. LSTMs are an extension of RNN with four neural networks in a single layer, where RNN have one only as shown in Fig. 2.9.

The main advantage of LSTM is the existence of state cell which is the line passing through in the top of every layer. The cell accounts for propagating information from previous to next layer. Then, "gates" in LSTM would manage which information will be passed or dropped. There are three gates to control the information flow, namely input, forget and output gates (Kim et al. 2016). These gates are composed of a sigmoid neural network and an operator as shown in Fig. 2.9.

2.2.7 K-Means Clustering

K-means clustering algorithm groups all observations data into k clusters iteratively until convergence will be reached. In the end, one cluster contains similar data since each data enters to the nearest cluster. K-means algorithm assigns a mean value of the cluster members as a cluster centroid. In every iteration, it calculates the shortest Euclidean distance from an observation data into any cluster centroid. Besides that, the intra-variances inside the cluster are also minimized by updating the cluster centroid iteratively. The algorithm would terminate when convergence is achieved, which the new clusters are the same as the previous iteration clusters (Jiang et al. 2017).

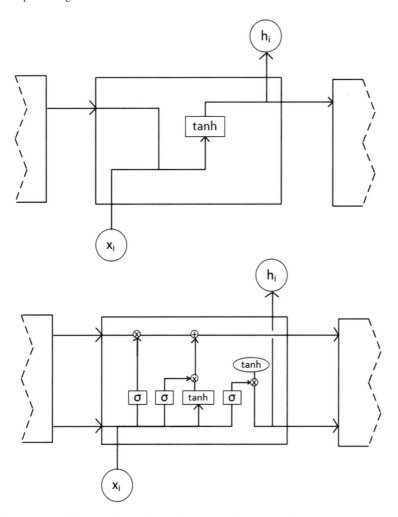

Fig. 2.9 RNN topology (Top) versus LSTM topology (Bottom) (Olah 2015)

2.2.8 Reinforcement Learning

Reinforcement Learning (RL) adopts a different approach to supervised and unsupervised learning. In RL, an agent assigned to take the responsibility of target classes in supervised learning which provides the correct corresponding class for each instance. The agent is responsible for deciding which action to perform its task (Olah 2017). Since no training data involved, the agent learns from experience during RL training. The learning process adopts trial and error approach during training to achieve the goal which is earn a long-term and highest reward. RL training often illustrated as a way for the player to win a game with some small points along the way and an ultimate prize at the end of the way. The player (or the RL agent) will explore ways to reach the final award. Sometimes, the agent will stuck in a small point due to its exploiting the minor points as its goal. Therefore, in RL, the exploration of new ways is needed although small points are reached. By this, the agent can achieve the ultimate prize at the end. In other words, the agent and game environment form a cyclic path of information which the agent do action, and the environment provides feedback based on the corresponding action. The process of reaching the ultimate prize can be formalized using Markov Decision Processes (MDP). In MDP, we can model the transition probabilities for each state. The objective function now is well represented by using MDP. There are two standard solutions to achieve this objective functions, Q-learning and policy learning. The former learns based on action-value function, while the latter learns using policy function which is a mapping between best action and corresponding state. More detail explanations are well delivered in Olah (2017).

In summary, Table 2.1 shows the commonly used layers in deep learning models.

Table 2.1 Commonly used layers in deep learning models

Deep learning layer		Description	Function
Activation function		ReLu	Maximum
		Sigmoid	Hyperbolic
		Tanh	Trigonometric
		Softmax	Hyperbolic
Pooling	Max pooling	Computing the maximum value of overlapping region in the preceding layer	Maximum
	Mean pooling	Computing the average value of non-overlapping region in the preceding layer	Mean
Fully connected		Dot product between the value of output neuron from the previous layer and the weight of the neuron	Matrix-vector multiplication
Dropout		Set random data to zero during training to prevent overfitting	Drop
Convolutional		Dot product between neighbor values in order to make convolution, then sum up the result	Weighted sum

References

Agrawal S, Boneh D, Boyen X (2010)Efficient lattice (H) IBE in the standard model. In: Annual international conference on the theory and applications of cryptographic techniques. Springer, pp 553–572

Boneh D, Franklin M (2001) Identity-based encryption from the weil pairing. In: Annual international cryptology conference. Springer, pp 213–229

Boneh D, Sahai A, Waters B (2011) Functional encryption: definitions and challenges. In: Theory of cryptography conference. Springer, pp 253–273

Boneh D, Waters B (2007) Conjunctive, subset, and range queries on encrypted data. In: Theory of cryptography conference. Springer, pp 535–554

Brakerski Z, Vaikuntanathan V (2014) Efficient fully homomorphic encryption from (standard) LWE. SIAM J Comput 43(2):831–871

Brakerski Z, Gentry C, Vaikuntanathan V (2014) (Leveled) fully homomorphic encryption without bootstrapping. ACM Trans Comput Theory (TOCT) 6(3):13

Brakerski Z, Vaikuntanathan V (2011) & #x201C;Fully homomorphic encryption from ring-LWE and security for key dependent messages. In: Advances in cryptology-CRYPTO. Springer, pp 505–524

Canetti R, Halevi S, Katz J (2003) A forward-secure public-key encryption scheme. In: International conference on the theory and applications of cryptographic techniques. Springer, pp 255–271

Chaum D (1983) Blind signatures for untraceable payments. In: Advances in cryptology. Springer, pp 199–203

Cheon JH, Coron J-S, Kim J, Lee MS, Lepoint T, Tibouchi M, Yun A (2013) Batch fully homomorphic encryption over the integers. In: Annual international conference on the theory and applications of cryptographic techniques. Springer, pp 315–335

Cheon JH, Kim A, Kim M, Song Y (2017) Homomorphic encryption for arithmetic of approximate numbers. In: International conference on the theory and application of cryptology and information security. Springer, pp 409–437

Chillotti I, Gama N, Georgieva M, Izabachène M (2020) TFHE: fast fully homomorphic encryption over the torus. J Cryptol 33(1):34–91

Chillotti I, Gama N, Georgieva M, Izabachene M (2016) Faster fully homomorphic encryption: Bootstrapping in less than 0.1 seconds. In: International conference on the theory and application of cryptology and information security. Springer, pp 3–33

Clear M, Goldrick CM (2017) Attribute-based fully homomorphic encryption with a bounded number of inputs. Int J Appl Cryptogr 3(4):363–376

Doweck J, Kao W-F, Lu AK-Y, Mandelblat J, Rahatekar A, Rappoport L, Rotem E, Yasin A, Yoaz A (2017) Inside 6th-generation intel core: new microarchitecture code-named skylake. IEEE Micro 37(2):52–62

Ducas L, Micciancio D (2015) FHEW: bootstrapping homomorphic encryption in less than a second. In: Annual International conference on the theory and applications of cryptographic techniques. Springer, pp 617–640

Dwork C, McSherry F, Nissim K, Smith A (2006) Calibrating noise to sensitivity in private data analysis. In: Theory of cryptography conference. Springer, pp 265–284

Fan J, Vercauteren F (2012) Somewhat practical fully homomorphic encryption. IACR Cryptol ePrint Arch 2012:144

Gentry C (2006) Practical identity-based encryption without random oracles. In: Annual international conference on the theory and applications of cryptographic techniques. Springer, pp 445–464

Gentry C (2009) Fully homomorphic encryption using ideal lattices. In: Annual ACM on symposium on theory of computing. ACM, pp 169–178

Gentry C, Peikert C, Vaikuntanathan V (2008) Trapdoors for hard lattices and new cryptographic constructions. In: Annual ACM on symposium on theory of computing. ACM, pp 197–206

Gentry C, Sahai A, Waters B (2013) & #x201C;Homomorphic encryption from learning with errors: conceptually-simpler, asymptotically-faster, attribute-based. In: Advances in cryptology-CRYPTO. Springer, pp 75–92

Goldreich O,Micali S, A Wigderson (1987) How to play ANY mental game. In: Proceedings of the nineteenth annual ACM symposium on theory of computing. pp 218–229

Goodfellow I, Pouget-Abadie J, Mirza M, Xu B, Warde-Farley D, Ozair S, Courville A, Bengio Y (2014) Generative adversarial nets. In: Advances in neural information processing systems. pp 2672–2680

Goyal V, Pandey O, Sahai A, Waters B (2006) Attribute-based encryption for fine-grained access control of encrypted data. In: Proceedings of the 13th ACM conference on computer and communications security. pp 89–98

Guyon I, Weston J, Barnhill S, Vapnik V (2002) Gene selection for cancer classification using support vector machines. Mach Learn 46(1–3):389–422

Halevi S, Shoup V (2014)Algorithms in HElib. In: International cryptology conference. Springer, pp 554–571

Hochreiter S, Schmidhuber J (1997) Long short-term memory. Neural Comput 9(8):1735–1780

Hunt T, Zhu Z, Xu Y, Peter S, Witchel E (2018) Ryoan: a distributed sandbox for untrusted computation on secret data. ACM Trans Comput Syst (TOCS) 35(4):1–32

Hunt T, Song C, Shokri R, Shmatikov V, Witchel E (2018) Chiron: privacy-preserving machine learning as a service. arXiv:1803.05961

Intel R (2014) Software guard extensions programming reference. Intel Corporation

Jiang C, Zhang H, Ren Y, Han Z, Chen K-C, Hanzo L (2017) Machine learning paradigms for next-generation wireless networks. IEEE Wirel Commun 24(2):98–105

Kim J, Kim J, Thu HLT, Kim H (2016) Long short term memory recurrent neural network classifier for intrusion detection. In: 2016 International conference on platform technology and service (PlatCon). IEEE, pp 1–5

LeCun Y, Haffner P, Bottou L, Bengio Y (1999) Object recognition with gradient-based learning. In: Shape, contour and grouping in computer vision. Springer, pp 319–345

Li N, Li T, Venkatasubramanian S (2007) t-closeness: privacy beyond k-anonymity and l-diversity. In: 2007 IEEE 23rd international conference on data engineering. IEEE, pp 106–115

Machanavajjhala A, Kifer D, Gehrke J, Venkitasubramaniam M (2007) l-diversity: privacy beyond k-anonymity. ACM Trans Knowl Discov Data (TKDD) 1(1):3–es

Olah C (2015) Understanding LSTM networks. http://colah.github.io/posts/2015-08-Understanding-LSTMs/. Accessed 20 Feb 2018

Olah C (2017) Machine learning for humans. https://www.dropbox.com/s/e38nil1dnl7481q/machine_learning.pdf?dl=0. Accessed 21 Mar 2018

Paillier P (1999) Public-key cryptosystems based on composite degree residuosity classes. In: International conference on the theory and applications of cryptographic techniques. Springer, pp 223–238

Qi L, Zhang X, Li S, Wan S, Wen Y, Gong W (2020) Spatial-temporal data-driven service recommendation with privacy-preservation. Inf Sci 515:91–102

Rivest RL, Adleman L, Dertouzos ML (1978) On data banks and privacy homomorphisms. Found Secur Comput 4(11):169–180

Sahai A, Waters B (2005) Fuzzy identity-based encryption. In: Annual international conference on the theory and applications of cryptographic techniques. Springer, pp 457–473

Samarati P, Sweeney L (1998) Protecting privacy when disclosing information: k-anonymity and its enforcement through generalization and suppression

Shamir A (1984) Identity-based cryptosystems and signature schemes. In: Workshop on the theory and application of cryptographic techniques. Springer, pp 47–53

Staudemeyer RC (2015) Applying long short-term memory recurrent neural networks to intrusion detection. South Afr Comput J 56(1):136–154

Van Dijk M, Gentry C, Halevi S, Vaikuntanathan V (2010) Fully homomorphic encryption over the integers. In: Annual international conference on the theory and applications of cryptographic techniques. Springer, pp 24–43

Waters B (2005) Efficient identity-based encryption without random oracles. In: Annual international conference on the theory and applications of cryptographic techniques. Springer, pp 114–127

Xiao X, Tao Y (2007) M-invariance: towards privacy preserving re-publication of dynamic datasets. In: Proceedings of the 2007 ACM SIGMOD international conference on management of data. pp 689–700

Yao AC-C, #x201C; (1986) How to generate and exchange secrets. In: Foundations of computer science, 27th Annual Symposium on. IEEE 1986:162–167

Zeng X, Chen Y-W, Tao C, van Alphen D (2009) Feature selection using recursive feature elimination for handwritten digit recognition. In: Proceedings of the intelligent information hiding and multimedia signal processing (IIH-MSP), Kyoto, Japan. IEEE, pp 1205–1208

Chapter 3
X-Based PPDL

Abstract We survey the latest publications regarding X-based privacy-preserving deep learning methods based on homomorphic encryption, secure multi-party computation, differential privacy, secure enclaves and their hybrid and summarize the key features of all surveyed publications including learning type and dataset.

3.1 HE-Based PPDL

HE-based PPDL combines HE with deep learning. The structure of HE-based PPDL is shown in Fig. 3.1. Generally, there are three phases in HE-based PPDL: the training phase (T1-T2-T3-T4), inference phase (I1-I2-I3), and result phase (R1-R2-R3). In the training phase, a client encrypts the training dataset using HE (T1) and sends the encrypted dataset to the cloud server (T2). In the cloud server, secure training is executed (T3), resulting in a trained model (T4). This is the end of the training phase. For the inference phase, the client sends the testing dataset to the cloud server (I1). The testing dataset becomes the input of the trained model (I2). Then, the prediction process is run using the trained model (I3), resulting in an encrypted computation result. This is the end of the inference phase. Next, the cloud server prepares to transport the encrypted computation result (R1) and sends it to the client (R2). The client finally decrypts it and obtains its computation result (R3).

Related Publications in 2016

Cryptonets was proposed by Gilad-Bachrach et al. (2016) to address the privacy issue in Machine Learning as a Service (MLaaS). The author combined cryptography and machine learning to present a machine learning framework that can receive encrypted data as an input. Cryptonets improves the performance of ML Confidential (2012) developed by Graepel et al., a modified PPDL scheme based on Linear Means Classifier (2002) and Fisher Linear Discriminant (2006) that works on HE. ML Confidential uses polynomial approximation to substitute for the nonlinear activation function. In this case, the PoM is not guaranteed because the client must generate

SpringerBriefs on Cyber Security Systems and Networks,
https://doi.org/10.1007/978-981-16-3764-3_3

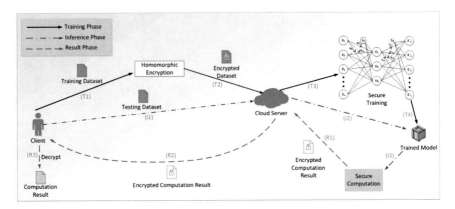

Fig. 3.1 The structure of HE-based PPDL

the encryption parameter based on the model. ML Confidential uses a cloud service-based scenario, and its main feature is ensuring the privacy of data during the transfer period between the client and the server. At first, the cloud server produces a public key and its private key for each client. Then, the client data are encrypted using HE and transferred to the server. The cloud server will perform the training process using the encrypted data and use the training model to perform classification on the testing dataset.

Cryptonets applies prediction based on encrypted data and then provides the prediction result, also in encrypted form, to users. Later, users can use their private key to decrypt the prediction result. By doing this, the privacy of the client and the privacy of the result are guaranteed. However, the privacy of model is not guaranteed because the client must generate an encryption parameter based on the model. The weakness of Cryptonets is the performance limitation because of the complexity issue. It does not work well on deeper NNs that have a large number of non-linear layers. In this case, the accuracy will decrease and the error rate will increase.

Cryptonets has trade-off between accuracy and privacy. This is caused by the utilization of activation function approximation using low-degree polynomial during the training phase. The neural network needs to be retrained again using plaintext with the same activation function in order to achieve good accuracy. Another weakness of Cryptonets is the limited number of neural network layer. The multiplicative leveled HE cannot be run on deep neural network with many layers. Faster Cryptonets (2018) accelerates homomorphic evaluation in Cryptonets (2016) by pruning network parameter such that many multiplication operations can be omitted. The main weakness of Faster Cryptonets is that it has vulnerability to membership inference attack (Fredrikson et al. 2015) and model stealing attack (Tramèr et al. 2016).

Related Publications in 2017

Aono17 (2017) is a PPDL system based on a simple NN structure. The author shows a weakness in the paper by Shokri and Shmatikov (2015) that leaks client data during the training process. The weakness is called Gradients Leak Information. It is an adversarial method for obtaining input values by calculating the gradient of the corresponding truth function to weight and the gradient of the corresponding of truth function to bias. If we divide the two results, we obtain the input value. Because of that reason, Aono17 (2017) proposes a revised PPDL method to overcome this weakness. The key idea is allowing the cloud server to update the deep learning model by accumulating gradient values from users. The author also utilized additively HE to protect gradient values against curious servers. However, a weakness actually remains in this approach because it does not prevent attacks between participants. Proper authentication of participants should be performed by the cloud server to prevent this vulnerability. This method is able to prevent data leakage by encrypting the gradient value. However, it has some limitations as the HE is compatible with parameter server only.

Chabanne17 (2017) is a PP scheme on DNN. The scheme is a combination of HE and CNN. The main idea is to combine Cryptonets (2016) with polynominal approximation for the activation function and batch normalization layer proposed by Ioffe and Szegedy (2015). The scheme wants to improve the performance of Cryptonets, which is only good when the number of non-linear layers in the model is small. The main idea is to change the structure of the regular NN by adding a batch normalization layer between the pooling layer and activation layer. Max pooling is not a linear function. As a result, in pooling layers average pooling is used instead of max pooling to provide the homomorphic part with a linear function. The batch normalization layer contributes to restricting the input of each activation layer, resulting in a stable distribution. Polynomial approximation with a low degree gives a small error, which is very suitable for use in this model. The training phase is performed using the regular activation function, and the testing phase is performed using the polynomial approximation as a substitution to non-linear activation function. Chabanne17 showed that their model achieved 99.30% accuracy, which is better than that of Cryptonets (98.95%). The pros of this model is its ability to work in a NN with a high number of non-linear layers while still providing higher than 99% accuracy, unlike Cryptonets which exhibits a decrease in accuracy when the number of non-linear layers is increased. Chabanne17's weakness is that the classification accuracy relies on the approximation of activation function. If the approximation function has high degree, it will be hard to get best approximation so that the accuracy will decrease.

CryptoDL (2017), proposed by Hesamifard et al., is a modified CNN for encrypted data. The activation function part of CNN is substituted with a low-degree polynomial. That paper showed that the polynomial approximation is indispensable for NN in HE environments. The authors tried to approximate three kinds of activation functions: ReLU, sigmoid, and tanh. The approximation technique is based on the derivative of the activation function. First, during the training phase, CNN with polynomial approximation is used. Then, the model produced during the training phase is used to perform classification over encrypted data. The authors applied the CryptoDL scheme to the MNIST dataset and achieved 99.52% accuracy. The weakness of this scheme is not covering PP training in DNN. The PP is only applied for the classification process. The advantage of this work is that it can classify many instances (8,192 or larger) for each prediction round, whereas DeepSecure (2018) classifies one instance per round. Hence, we can say that CryptoDL works more effectively than DeepSecure. The weakness of CryptoDL is claimed to be the limited number of layers in DNN. Since as the number of layer increases, the complexity is also increased multiplicatively due to HE operations, reducing its performance like Cryptonets (2016).

Related Publications in 2018

In TAPAS (2018), the author addresses the weakness of FHE in PPDL, which requires a large amount of time to evaluate deep learning models for encrypted data (Chillotti et al. 2016). The author developed a deep learning architecture that consists of a fully-connected layer, a convolutional layer, and a batch normalized layer (Ioffe and Szegedy 2015) with sparsified encrypted computation to reduce the computation time. The main contribution here is a new algorithm to accelerate binary computation in the binary neural network (Kim and Smaragdis 2016; Hubara et al. 2017). Another superiority of TAPAS is supporting parallel computing. The technique can be parallelized by evaluating gates in the same level at the same time. A serious limitation of TAPAS is that it only supports binary neural network. In order to overcome this limitation, a method to encrypt non-binary or real-valued neural network is required.

FHE DiNN (2018) is a PPDL framework that combines FHE with a discretized neural network. It addresses the complexity problem of HE in PPDL. FHE-DiNN offers a NN with linear complexity with regard to the depth of the network. In other words, FHE-DiNN has the scale invariance property. Linearity is achieved by the bootstrapping procedure on a discretized NN with a weighted sum and a sign activation function that has a value between -1 and 1. The sign activation function will maintain linearity growth such that it will not be out of control. The computation of the activation function will be performed during the bootstrapping procedure to refresh the ciphertext, reducing its cumulative noise. When we compare the discretized neural network to a standard NN, there is one main difference: the weight, the bias value, and the domain of the activation function in FHE DiNN needs to be discretized. The sign activation function is used to limit the growth of the signal in the range of -1 and 1, showing its characteristic of linear scale invariance for linear complexity. Compared with Cryptonets (2016), FHE DiNN successfully improves the speed and reduces the complexity of FHE but with a decrease in accuracy; thus, a trade-off exists. The weakness of this method happens in the discretization process, which uses sign activation function that leads to a decrease in accuracy. It gets better if the training process is directly executed in a discretized neural network, rather than by converting a regular network into a discretized one.

E2DM (2018) converts an image dataset into matrices. The main purpose of doing this is to reduce the computational complexity. E2DM shows how to encrypt multiple matrices into a single ciphertext. It extends some basic matrix operations such as rectangular multiplication and transposition for advanced operations. Not only is the data encrypted; the model is also homomorphically encrypted. As a result, PoC and PoM are guaranteed. E2DM also fulfills the PoR as only the client can decrypt the prediction result. For the deep learning part, E2DM utilizes CNN with one convolutional layer, two fully connected layers, and a square activation function. The weakness of E2DM is that it can only support simple matrix operation. Extending the advanced matrix computation will be a promising future work.

Xue18 (2018) tries to enhance the scalability of the current PPDL method. A PPDL framework with multi-key HE was proposed. Its main purpose (Xue et al. 2018) was to provide a service to classify large-scale distributed data. For example, in the case of predicting road conditions, the NN model must be trained from traffic information data from many drivers. For the deep learning structure, (Xue et al. 2018) modification to the conventional CNN architecture is necessary, such as changing max pooling into average pooling, adding a batch normalization layer before each activation function layer, and replacing The ReLU activation function with a low-degree approximation polynomial. PoC and PoR are guaranteed here. However, the privacy of the model is not guaranteed because the client must generate an encryption parameter based on the model. The weakness of this approach is that the neural network must be trained by using encrypted data during the training phase. So, privacy leakage may happen if appropriate countermeasure is not deployed.

Liu18 (2018) is a PP technique for convolutional networks using HE. The technique uses an MNIST dataset that contains handwritten numbers. Liu18 (2018) encrypts the data using HE and then uses the encrypted data to train CNN. Later, the classification and testing process is performed using the model from CNN. The idea is to add a batch normalization layer before each activation layer and approximate activation layer using Gaussian distribution and the Taylor series. The non-linear pooling layer is substituted for with the convolutional layer with increased stride. By doing this, the author successfully modified CNN to be compatible with HE, achieving 98.97% accuracy during the testing phase. The main difference between regular CNN and modified CNN in PP technology is the addition of the batch normalization layer and the change of the non-linear function in the activation layer and the pooling layer into a linear function. The proposed approach has weakness from the point of complexity since the HE has massive computational overhead leading to huge memory overhead.

Related Publications in 2019

CryptoNN (2019) is a PP method that utilizes functional encryption for arithmetic computation over encrypted data. The FE scheme protects the data in the shape of a feature vector inside matrices. By doing this, the matrix computation for NN training can be performed in encrypted form. The training phase of CryptoNN comprises two main steps: a secure feed-forward step and a secure back-propagation step. The CNN model is adapted with five main functions: a dot-product function, weighted-sum function, pooling function, activation function, and cost function. During the feed-forward phase, the multiplication of the weight value and feature vector cannot be performed directly because the vector value is encrypted. As a result, a function-derived key is used to transform the weight value such that it can computed. However, the scalability of CryptoNN is still in question since the dataset used in their experiment is a simple one. It needs to be tested with more complex dataset and deeper neural network model.

Zhang19 (2019) is a secure clustering method for preserving data privacy in cloud computing. The method combines a probabilistic C-Means algorithm (Krishnapuram and Keller 1996) with a BGV encryption scheme (Hesamifard et al. 2017) to produce HE-based big data clustering on a cloud environment. The main reason for choosing BGV in this scheme is its ability to ensure a correct result on the computation of encrypted data. The author also addresses the weakness of the probabilistic C-Means algorithm, which is very sensitive and needs to be initialized properly. To solve this problem, fuzzy clustering (Gustafson and Kessel 1978) and probabilistic clustering (Smyth 2000) are combined. During the training process, there are two main steps: calculating the weight value and updating the matrix. To this end, a Taylor approximation for the activation function is used as the function is polynomial with addition and multiplication operations only. The main weakness is that the computation cost will increase proportionally to the number of neural network layers due to characteristic of HE.

Related Publications in 2020

Saerom20 (2020) proposed an HE algorithm for machine learning training using Support Vector Machine (SVM). This work is the first practical method that implements HE on SVM. The experiment result showed that the proposed approach successfully outperform the current logistic regression classifier on real world dataset. They successfully address the constrained optimization problem that appears because of inefficient operation in an encrypted field by introducing the least square SVM algorithm. The least square SVM algorithm is based on gradient descent theory for least square problem. Another contribution of the proposed method is adaptable to multi-class classification problem by implementing parallel calculation.

CryptoRNN (2020) proposed a PP framework using HE on Recurrent Neural Networks (RNN). The proposed method successfully implemented encrypted learning on RNN, while addressing two main issues: noise increase in encrypted operation and the compatibility of activation function in HE operation. They handle the noise increase issue by performing a non-linear calculation when the server sends ciphertexts to the client and refreshing the ciphertexts after each multiplication process to reduce the size of ciphertexts and make the inference faster. They also used a polynomial activation function that refresh the ciphertexts after each round of classification to reduce the communication cost with the client.

Jaehyoung20 (2020) proposed a Privacy-Preserving Reinforcement Learning (PPRL) framework for cloud computing service. The proposed method using FHE based on Learning With Errors (LWE). PPRL is more efficient than SMC since all data are stored and processed in a single cloud server, resulting in less communication overhead. They also discarding the bootstrapping algorithm in FHE and substitute it with iterative Q-value computation. They successfully cancelled the error growth using this method. The confidentiality of the user is also guaranteed by applying RSA encryption during the data exchange phase. The experiments showed that the proposed method has significantly less communication overhead than regular SMC-based PP algorithm.

Table 3.1 shows the Key features of HE-based PPDL methods using (Boulemtafes et al. 2020) classification for the role of edge server. They divided PPDL techniques into two kinds; server-based and server-assisted. Server-based means that the learning process is executed on the cloud server. On the other hand, server-assisted means that the learning process is performed collaboratively by the parties and the server.

3.2 Secure MPC-Based PPDL

Generally, the structure of a secure MPC-based PPDL is shown in Fig. 3.2. Firstly, users perform local training using their private data (1). Then, the gradient result from the training process is secret-shared (2). The shared gradient is transmitted to each server (3). After that, the server aggregates the shared gradient value from users (4). The aggregated gradient value is transmitted from each server to each client (5). Each client reconstructs the aggregated gradient and updates the gradient value for the next training process (6). In the case of multi-party computation, secret sharing is used to preserve the data privacy. However, for specific secure two-party computation, a garbled circuit with secret sharing is widely used instead of secret sharing.

The structure of secure two-party computation is shown in Fig. 3.3. In secure two-party computation, a client uses garbled circuit to protect the data privacy. The communication between the client and the server is securely guaranteed by using oblivious transfer. At first, a client sends the private data input to the garbled circuit for the garbling process (1). Then, the next process is the data exchange between the client and the server using oblivious transfer (2). After the data exchange has been completed, the server runs the prediction process, using the data as an input in the deep learning model (3). The prediction result is sent back to the client. The client uses the garbled table to aggregate the result (4) and obtain the final output (5).

Related Publications in 2017

SecureML (2017) is a new protocol for privacy-preserving machine learning. The protocol uses Oblivious Transfer (OT), Yao's GC, and Secret Sharing to ensure the privacy of the system. For the deep learning part, it leverages linear regression

Table 3.1 Key features of HE-based PPDL methods

References	Key concept	Learning type	Dataset
Cryptonets (2016)	Enables cloud-based NN training using polynomial approximation of activation function	Server-based	MNIST
Aono17 (2017)	Enables collaborative learning between participants over combined datasets	Server-assisted	MNIST
Chabanne17 (2017)	Applies FHE with low degree approximation of activation function	Server-based	MNIST
CryptoDL (2017)	Applies leveled HE with approximation of activation function based on the derivative of the function	Server-based	MNIST CIFAR-10
TAPAS (2018)	Proposes a sparsified encrypted computation to speed up the computation in binary neural network	Server-based	MNIST
FHEDiNN (2018)	Applies FHE in discretized neural network with linear complexity, in regards to the network's depth	Server-based	MNIST
E2DM (2018)	Proposes multiple matrices encryption into a single chipertext to reduce computational complexity	Server-based	MNIST
Xue18 (2018)	Applies multi-key HE with batch normalization layer before each activation function layer	Server-based	MNIST
Liu18 (2018)	Proposes the addition of batch normalization layer and approximates activation function using Gaussian distribution and Taylor series	Server-based	MNIST
Faster Cryptonets (2018)	Accelerates homomorphic evaluation by pruning the network parameters	Server-based	MNIST
CryptoNN (2019)	Utilizes FE for arithmetic computation over encrypted data	Server-based	MNIST
Zhang19 (2019)	Combines probabilistic C-Means algorithm with BGV encryption for big data clustering in cloud environment	Server-based	eGSAD Swsn
Saerom20 (2020)	Combines HE with SVM using least square gradient descent theory	Server-based	UCI
CryptoRNN (2020)	Combines HE with RNN and reduce the noise increase by refreshing the ciphertext each round	Server-based	UCI
Jaehyoung20 (2020)	Combines HE with RL using LWE and reducing the complexity by substituting bootstrapping procedure with iterative Q-value computation	Server-based	–

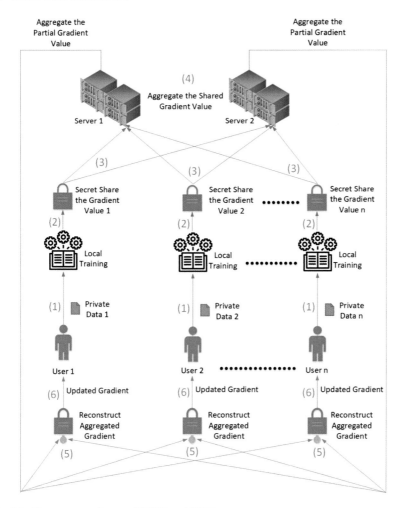

Fig. 3.2   The structure of secure MPC-Based PPDL

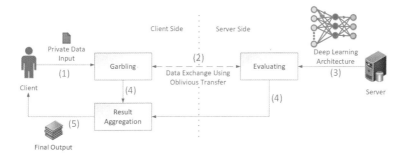

Fig. 3.3   The structure of secure two-party computation-based PPDL

and logistic regression in a DNN environment. The protocol proposes an addition and multiplication algorithm for secretly shared values in the linear regression. The Stochastic Gradient Descent (SGD) method is utilized to calculate the optimum value of regression. The weakness of this scheme is that it can only implement a simple NN without any convolutional layer; thus, the accuracy is quite low. The weakness of SecureML relies on the non-colluding assumption. In the two-servers model, the servers can be untrusted but not collude with each other. If the servers may collude, the privacy of participants can be compromised.

MiniONN (2017) is a PP framework for transforming a NN into an oblivious Neural Network. The transformation process in MiniONN includes the nonlinear functions, with a price of negligible accuracy lost. There are two kinds of transformation provided by MiniONN, including oblivious transformation for the piecewise linear activation function and oblivious transformation for the smooth activation function. A smooth function can be transformed into a continuous polynomial by splitting the function into several parts. Then, for each part, polynomial approximation is used for the approximation, resulting in a piecewise linear function. Hence, MiniONN supports all activation functions that have either a monotonic range or piecewise polynomial or can be approximated into a polynomial function. The experiment showed that MiniONN outperforms Cryptonets (2016) and SecureML (2017) in terms of message size and latency. The main weakness is that MiniONN does not support batch processing. MiniONN is also based on honest-but-curious adversary, so it has no countermeasure against malicious adversary.

Related Publications in 2018

ABY3 (2018) proposed by Mohassel et al., is a protocol for privacy-preserving machine learning based on three-party computation (3PC). The main contribution of this protocol is its ability to switch among arithmetic, binary, and Yao's 3PC depending on the processing needs. The main purpose of ABY3 is to solve the classic PPDL problem that requires switching back and forth between arithmetic (for example addition and multiplication) and non-arithmetic operations (such as activation function approximation). The usual machine learning process works on arithmetic operations. As a result, it cannot perform a polynomial approximation for activation function. ABY3 can be used to train linear regression, logistic regression, and NN models. Arithmetic sharing is used when training linear regression models. On the other hand, for computing logistic regression and NN models, binary sharing on three-party GC is utilized. The author also introduced a new fixed-point multiplication method for more than three-party computation, extending the 3PC scenario. This multiplication method is used to solve the limitation of using MPC with machine learning. MPC is suitable for working over rings, unlike machine learning which works on decimal values. ABY3 provides a new framework that is secure against malicious adversaries; so it is not limited to honest-but-curious adversary. However, since the protocols are built in their own framework, it will be difficult to be implemented with other deep learning scheme.

DeepSecure (2018) is a framework that enables the use of deep learning in PP environments. The author used OT and Yao's GC protocol (1986) with CNN to perform the learning process. DeepSecure enables a collaboration between client and server to perform the learning process on a cloud server using data from the client. The security of the system was proven using an honest-but-curious adversary model. The GC protocol successfully keeps the client data private during the data transfer period. The weakness of this method is its limitation of number of instances processed each round. The method can only classify one instance during each prediction round. DeepSecure offers a preprocessing phase that reduces the size of data. The strength of DeepSecure is that the preprocessing phase can be adopted easily because it is independent from any cryptographic protocol. Its main weakness is the inability to process batch processing.

Chameleon (2018) is a PPDL method that combines Secure-MPC and CNN. For the privacy part, Chameleon uses Yao's GC which enables two parties to perform joint computation without disclosing their own input. There are two phases: an online phase and offline phase. During the online phase all parties are allowed to communicate, whereas during the offline phase the cryptographic operations are precomputed. Chameleon utilizes vector multiplication (dot product) of signed fixed-point representation which improves the efficiency of heavy matrix multiplication for encrypted data classification. It successfully achieves faster execution compared with CryptoNets (2016) and MiniONN (2017). Chameleon requires two non-colluding servers to ensure the data privacy and security. For the private inference, it requires an independent third party or a secure hardware such as Intel SGX. Chameleon is based on honest-but-curious adversary, there is no countermeasure against malicious adversary. Chameleon's protocol is based on two party computation, so it is not efficient to implement in more than two-party scenario.

Related Publications in 2019

SecureNN (2019) provides the first system that ensures the privacy and correctness against honest-but-curious adversaries and malicious adversaries for complex NN computation. The system is based on secure MPC combined with CNN. SecureNN was tested on an MNIST dataset and successfully achieved more than 99% prediction accuracy with execution times 2–4 times faster than other secure MPC based PPDL, such as SecureML (2017), MiniONN (2017), Chameleon (2018), and GAZELLE (2018). Its main contribution is developing a new protocol for Boolean computation (ReLU, Maxpool, and its derivatives) that has less communication overhead than Yao GC. This is how SecureNN achieves a faster execution time than the other techniques mentioned above. The weakness of SecureNN is claimed to refine more communication overhead compared to ABY3 [86]. If the SecureNN protocol is modified so that it utilizes matrix multiplication like ABY3, the number of communication rounds will be reduced.

CodedPrivateML (2019) distributes the training computation across several stations and proposes a new approach for secret sharing of the data and DL model parameter that significantly reduces the computation overhead and complexity. However, the accuracy of this method is only about 95%, which is not as high as other method such as GAZELLE (2018) or Chameleon (2018).

Related Publications in 2020

Tran20 (2021) proposed a decentralized secure MPC framework for PP. Their proposed system is able to ensure the privacy of all parties with low communication cost called as Efficient Secure Sum Protocol (ESSP) that enables a secure joint computation without needing a trusted third-party server. Their protocol can work with floating point data type and support parallel training process, which looks like an advantage. Their experiment showed that ESSP protocol can protect against honest-but-curious parties (maximum n-2 of n parties colluding). Their approach also has good accuracy, achieving 97% baseline accuracy with less round of training compared to Downpour Stochastic Gradient Descent (2012) technique.

Ramirez20 (2020) proposed a kind of K-means clustering for secure MPC computation. Their method combines SecureNN (2019) with additive secret sharing and extends the protocol over horizontal and vertical data partitionings. They designed an algorithm to do horizontal data partition for grouping data points while keeping the privacy and proposed a secure vertical k-means algorithm that allows local computation over vertically partitioned data across parties.

Liu20 (2020) proposed a distributed secure MPC framework for privacy-preserving data mining. Their framework is designed based on SPDZ protocol (2020), combined with one-hot-encoding and lower-upper decomposition algorithms. Their algorithm supports variable encoding, data vectorization, and matrix operation. Lower-upper decomposition algorithm (Schwarzenberg-Czerny 1995) is chosen to solve regression problem with less time-consuming operation. The experiment showed that the proposed method is feasible and effective based on preciseness, execution time, byte code size, and transmission data size metrics.

Table 3.2 shows the Key features of secure MPC-based PPDL methods.

3.3   Differential Privacy-Based PPDL

The structure of differential privacy-based PPDL is shown in Fig. 3.4. First, training data are used to train the teacher model (1). Then, the teacher model is used to train the student model. In this case we illustrated the student model as a GAN model that consists of a generator and discriminator (2). Random noise is added to the generator as it generates fake training data (3). On the other hand, the teacher model trains the student model using the public data (4). The student model runs a zero-sum game between the generator and the discriminator. Then, the student model is ready to

Table 3.2 Key features of secure MPC-based PPDL methods

References	Key concept	Learning type	Dataset
SecureML (2017)	Proposes a combination of garbled circuit with oblivious transfer and secret sharing in a DNN environment	Server-assisted	MNIST CIFAR-10
MiniONN (2017)	Transforms a NN into an oblivious NN	Server-assisted	MNIST
ABY3 (2018)	Provides an ability to switch between arithmetic, binary, and three-party computation	Server-assisted	MNIST
DeepSecure (2018)	Enables a collaboration between client and server to do a learning process on cloud server	Server-assisted	MNIST
Chameleon (2018)	Enables a secure joint computation with two distinguished phases; online and offline	Server-assisted	MNIST
SecureNN (2019)	Develops a new protocol for Boolean computation that has small overhead	Server-assisted	MNIST
Coded PrivateML (2019)	Proposes a distributed training computation across clients with a new approach of secret sharing	Server-assisted	MNIST
Tran20 (2021)	Proposes a decentralized computation without any need of a trusted third-party server. It also can work with floating point data type	Server-assisted	MNIST UCI
Ramirez20 (2020)	Proposes a K-means clustering algorithm for Secure MPC computation that can support horizontal and vertical data partitioning	Server-assisted	–
Liu20 (2020)	Proposes a distributed Secure MPC framework for privacy-preserving data mining with one-hot-encoding and lower-upper decomposition algorithm	Server-assisted	UCI

be used for the prediction process. A client sends a query (5) to the student model. The student model runs the inference phase and returns the prediction result to the user (6).

Private Aggregation of Teacher Ensembles (PATE) (2016) is a PPDL method for MLaaS that uses a differential privacy-based approach in Generative Adversarial Network (GAN). PATE is a black box approach that tries to ensure the privacy of data during training by using teacher-student models. During the training phase, the dataset is used to train the teacher models. Then, student models learn from

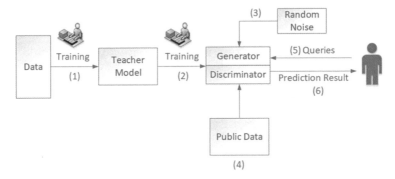

Fig. 3.4 The structure of differential Privacy-based PPDL

the teacher models using a voting-based differential privacy method. By doing this, the teacher model is kept secretive, and the original data cannot be accessed by the student. The advantage of this model is due to the distinguished teacher model; when an adversary obtains a student model, the model will not give the adversary any confidential information. PATE has a serious weakness, which is not to provide good accuracy for complex data. If the data is too diverse, adding noise to the data will lower the performance of PATE. So, the performance of PATE depends on the type of input. It is only suitable for simple classification task. Furthermore, the computation cost is expensive due to many interactions between server and clients.

Another PPDL method that utilizes differential privacy is Bu19 (2019). Bu19 proposes Gaussian Differential Privacy (Gaussian DP) which formalizes the original DP technique as a hypothesis test from the adversaries' perspective. The concept of adding Gaussian noise is interesting. It must be evaluated in order to analyze the trade-off between the noise and the accuracy. The scalability issue implemented in the daily life remains in question.

Gong20 (2020) proposed an Adaptive Differential Privacy Preserving Learning (ADPPL) framework in deep neural network. The framework is based on relevance analysis to address the gap between private and non-private models. It works by perturbing gradient between neurons in the layers and the model output. During the backpropagation process, Gaussian noise is added to gradient of neuron that have less relevance to the model. On the other hand, less noise is added to gradient of neuron that have more relevance to the model. Their experiment showed that the proposed method significantly improved the accuracy compared to traditional differential privacy approach.

Fan20 (2020) proposed a local differential privacy framework for data centers. The goal is to address the Laplace noise issue during the pattern mining process. In order to achieve that goal, they design an algorithm to quantify the quality of privacy protection. Their algorithm satisfies local differential privacy and guarantees the validity of privacy protection by adding Laplacian mechanism. The experiment result showed that their proposed algorithm successfully improved the efficiency, security, and accuracy of the classification process in the data center.

Bi20 (2020) proposed a location data collection method to satisfy differential privacy in edge computing. The method is based on Local Differential Privacy (LDP) model. They also utilized a road network space division method based on Voronoi diagram to perturb the original data. The disturbance on each Voronoi grid has been successfully proven to satisfy LDP. The experiment showed that their proposed method has successfully protected users' location privacy in edge computing.

In Table 3.3, the Key features of differential privacy-based PPDL methods are summarized.

Table 3.3 Key features of differential privacy-based PPDL methods

References	Key concept	Learning type	Dataset
PATE (2016)	Proposes a differentially private learning process by utilizing teacher models and student models	Server-based	MNIST SVHN
Bu19 (2019)	Proposes a Gaussian differential privacy that formalizes the original differential privacy-based PPDL	Server-based	MNIST MovieLens
Gong20 (2020)	Proposes an Adaptive Differential Privacy Preserving Learning (ADPPL) framework in deep neural network using relevance analysis	Server-based	MNIST CIFAR-10
Fan20 (2020)	Proposes a local differential privacy framework for data centers using Laplacian mechanism to measure the quality of privacy protection	Server-based	UCI
Bi20 (2020)	Proposes a location data collection method using a road network space division method based on Voronoi diagram to satisfy differential privacy in edge computing	Server-based	Gowalla

3.4   Secure Enclaves-Based PPDL

The structure of secure enclaves-based PPDL is shown in Fig. 3.5. At first, a client sends data to the secure enclave environment (1). Then, the model provider sends the deep learning model to the enclaves (2). In the secure enclaves environment, the prediction process is executed using the client's data and the deep learning model (3). Then, the prediction result is sent to the client (4). The process in secure enclaves is guaranteed to be secure, and all of the data and models inside cannot be revealed to any other party outside the enclaves.

SLALOM (2018) uses Trusted Execution Environments (TEEs), which isolate the computation process from untrusted software. The DNN computation is partitioned between trusted and untrusted parties. SLALOM runs DNN in the Intel SGX enclave which delegates the computation process to an untrusted GPU. The weakness of this approach is believed to limit CPU operation since the TEE does not allow to access GPU. A vulnerability by side channel attack may occur as shown by Van Bulck et al. (2018).

Chiron (2018) provides a black-box system for PPDL. The system conceals training data and model structure from the service provider. It utilizes SGX enclaves (2016) and the Ryoan sandbox (2018). As SGX enclaves only protect the privacy of the model, the Ryoan sandbox is chosen here to ensure, even if the model tries to leak the data, that the data will be confined inside the sandbox, preventing the leakage. Chiron also supports a distributed training process by executing multiple enclaves that exchange model parameters through the server.

Chiron focuses on outsourced learning by using a secure enclave environment. The main difference between Chiron and Ohrimenko16 (2016) is the code execution. Chiron allows the execution of untrusted code to update the model and implements protection by using sandboxes such that the code will not leak the data outside the

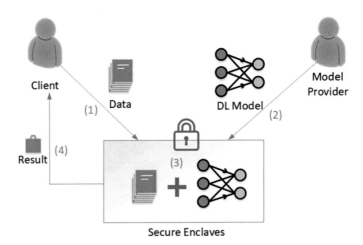

Fig. 3.5 The structure of secure Enclaves-based PPDL

enclave. On the other hand, Ohrimenko16 requires all codes inside the SGX enclave to be public to ensure that the code is trusted. The main weakness relies on the assumption that the model is not exposed to other parties. As a result, if an adversary can get an access to the trained model, there will be leakage of data, as shown by Shokri et al. (2017). This leakage problem can be solved by using differential privacy-based training algorithm (Papernot et al. 2016).

TrustAV (2020) is a cloud-based privacy-preserving malware detection framework in a secure enclaves environment. It works by offloading the malware analysis process to a remote server. The entire analysis is run in the server, inside a hardware supported secure enclaves, so it is totally safe. By doing this, it can preserve the data transfer in an untrusted environment with acceptable performance overhead. TrustAV has successfully ensured the data privacy against malicious adversaries and honest-but-curious provider. It also reduces the required enclaves memory in SGX technology up to three times better performance by implementing parameter limitation for signature-based solution. Their caching scheme, protection of signature-based automata, and reduced enclaves memory footprint are the main features of TrustAV.

Chen20 (2020) proposed a privacy-preserving federated learning scheme in a trusted execution environment to prevent causative attack that happens a type of common attacks in federated learning, where a malicious participant injects false training results in order to corrupt the training model. Their scheme has a robust protocol that can detect causative attack and protect the privacy of all participants. In a nutshell, the proposed method has successfully ensured the privacy and integrity of all participants. The training scheme enables collaborative learning with data privacy since participants can do training process using their own inputs, without revealing those inputs to other participants. The scheme also utilizes an algorithm to detect any dishonest action (tampering training model or delaying training process), so the participants' integrity is guaranteed.

Law et al. proposed Secure XGBoost (2020), a PP framework that enables collaborative training and inference in a trusted environment. The framework is based on the popular XGBoost library. Secure XGBoost utilizes hardware enclaves to ensure the privacy and integrity of all parties. The system augments a novel data-oblivious algorithm to prevent the side channel attacks by using access pattern leakage. The algorithm prevents attackers from monitoring enclaves' memory access pattern so that sensitive information can be protected. Their experiments showed that Secure XGBoost has successfully protected the privacy and integrity as an oblivious distributed solution for gradient boosted decision trees. The performance of Secure XGBoost is heavily depending on the hyper-parameters. Increasing the number of trees while decreasing the number of bins can reduce the overhead by maintaining the same accuracy.

Table 3.4 shows the Key features of secure enclaves-based PPDL methods.

Table 3.4 Key features of secure enclaves-based PPDL methods

References	Key concept	Learning type	Dataset
Chiron (2018)	Combines SGX enclaves and Ryoan sandbox to provide a black-box system for PPDL	Server-based	CIFAR ImageNet
SLALOM (2018)	Utilizes a trusted execution environment to isolate computation processes	Server-based	–
TrustAV (2020)	Proposes a cloud-based privacy-preserving malware detection framework in a secure enclaves environment with reduced required SGX memory	Server-based	–
Chen20 (2020)	Proposes a privacy-preserving federated learning scheme in a trusted execution environment to prevent causative attack	Server-based	MNIST Foursquare
Secure XGBoost (2020)	Proposes a privacy-preserving framework with data-oblivious algorithm that enables collaborative training and inference in a trusted environment	Server-based	Ant financial

3.5   Hybrid-Based PPDL

Ohrimenko16 (2016) proposes a secure enclave platform based on the SGX system for secure MPC. It focuses on collaborative learning, providing a prediction service in a cloud. Ohrimenko16 requires all codes inside the SGX enclave to be public to ensure that the code is trusted. The main weakness of this method is claimed to its inherent vulnerability to information leakage due to GAN attack as shown by Hitaj et al. (2017).

Chase17 (2017) wants to propose a private collaborative framework for machine learning. The main idea is to combine secure MPC with DP for the privacy part and leverage NN for the machine learning part. The weakness of this method is found to undergo a decrease in accuracy when implemented in a large network, exhibiting the scalability issue. In addition, its data privacy can only be guaranteed if the participants are non-colluding.

In GAZELLE (2018), HE is combined with GC to ensure privacy and security in a MLaaS environment. For the HE library, it utilizes Single Instruction Multiple Data (SIMD) which includes addition and multiplication of ciphertext to improve the encryption speed. The Gazelle algorithm accelerates the convolutional and the matrix multiplication processes. An automatic switch between HE and GC is implemented such that encrypted data can be processed in NN. For the deep learning part, it leverages CNN comprising two convolutional layers, two ReLU layers as activation layers, one pooling layer, and one fully connected layer. The author used MNIST and CIFAR-10 datasets during the experiment and successfully showed that Gazelle outperforms several popular techniques such as MiniONN (2017) and Cryptonets (2016) in terms of run time. Furthermore, to prevent a linkage attack, Gazelle limits the number of classification queries from a client. The limitation of GAZELLE is claimed to support two-party computation scheme only since it utilizes garbled circuit for the secure exchange of two parties.

Ryffel18 (2018) introduces a PPDL framework using federated learning built over PyTorch (2019). Federated learning requires multiple machines to train data in a decentralized environment. It enables clients to learn a shared prediction model using the data in their own device. The author combines secure MPC with DP to build a protocol enables federated learning. Overall, the proposed approach has overhead problem because of the bottleneck in the low-level library, compared to the high level python API. The proposed approach is vulnerable to collusion attack if the participants collude with each other.

CrypTFlow (2019) combines secure enclaves with secret sharing in DNN to secure the learning process of the ImageNet dataset. The main weakness of CrypTFlow is believed not to support GPU processing. As a result, the computation overhead during the secure training is still high.

Table 3.5 shows the key features of hybrid-based PPDL methods.

Table 3.5 Key features of hybrid-based PPDL methods

References	Key concept	Learning type	Dataset
Ohrimenko16 (2016)	Applies secure enclaves platform based for secure MPC	Server-based	MovieLens MNIST
Chase17 (2017)	Proposes a private collaboration framework by combining secure MPC with DP	Server-assisted	MNIST
GAZELLE (2018)	Combines HE with garbled circuit using SIMD	Server-based	MNIST CIFAR-10
Ryffel18 (2018)	Builds a new protocol for federated learning in PPDL	Server-assisted	Boston Housing Pima Diabetes
CrypTFlow (2019)	Combines secure enclaves with secret sharing in DNN	Server-assisted	ImageNet

References

Aono Y, Hayashi T, Wang L, Moriai S et al (2017) Privacy-preserving deep learning via additively homomorphic encryption. IEEE Trans Inf Forensics Secur 13(5):1333–1345

Bakshi M, Last M (2020) CryptoRNN-privacy-preserving recurrent neural networks using homomorphic encryption. In: International symposium on cyber security cryptography and machine learning. Springer, pp 245–253

Bi M, Wang Y, Cai Z, Tong X (2020) A privacy-preserving mechanism based on local differential privacy in edge computing. China Commun 17(9):50–65

Boulemtafes A, Derhab A, Challal Y (2020) A review of privacy-preserving techniques for deep learning. Neurocomputing 384:21–45

Bourse F, Minelli M, Minihold M, Paillier P (2018) Fast homomorphic evaluation of deep discretized neural networks. In: Annual international cryptology conference. Springer, pp 483–512

Bu Z, Dong J, Long Q, Su WJ (2019) Deep learning with gaussian differential privacy. arXiv:1911.11607

Chabanne H, de Wargny A, Milgram J, Morel C, Prouff E (2017) Privacy-preserving classification on deep neural network. IACR Cryptol Arch 2017:35

Chase M, Gilad-Bachrach R, Laine K, Lauter KE, Rindal P (2017) Private collaborative neural network learning. IACR Cryptol Arch 2017:762

Chen Y, Luo F, Li T, Xiang T, Liu Z, Li J (2020) A training-integrity privacy-preserving federated learning scheme with trusted execution environment. Inf Sci 522:69–79

Chillotti I, Gama N, Georgieva M, Izabachene M (2016) Faster fully homomorphic encryption: Bootstrapping in less than 0.1 s. In: International conference on the theory and application of cryptology and information security. Springer, pp 3–33

Chou E, Beal J, Levy D, Yeung S, Haque A, Fei-Fei L (2018) Faster cryptonets: leveraging sparsity for real-world encrypted inference. arXiv:1811.09953

Dean J, Corrado G, Monga R, Chen K, Devin M, Mao M, Ranzato M, Senior A, Tucker P, Yang K et al (2012) Large scale distributed deep networks. In: Advances in neural information processing systems, pp 1223–1231

Deyannis D, Papadogiannaki E, Kalivianakis G, Vasiliadis G, Ioannidis S (2020) TrustAV: practical and privacy preserving malware analysis in the cloud. In: Proceedings of the tenth ACM conference on data and application security and privacy, pp 39–48

Fan W, He J, Guo M, Li P, Han Z, Wang R (2020) Privacy preserving classification on local differential privacy in data centers. J Parallel Distrib Comput 135:70–82

Fredrikson M, Jha S, Ristenpart T (2015) Model inversion attacks that exploit confidence information and basic countermeasures. In: Proceedings of the 22nd ACM SIGSAC conference on computer and communications security, pp 1322–1333

Gilad-Bachrach R, Dowlin N, Laine K, Lauter K, Naehrig M, Wernsing J (2016) Cryptonets: applying neural networks to encrypted data with high throughput and accuracy. In: International conference on machine learning, pp 201–210

Gong M, Pan K, Xie Y, Qin AK, Tang Z (2020) Preserving differential privacy in deep neural networks with relevance-based adaptive noise imposition. Neural Netw 125:131–141

Graepel T, Lauter K, Naehrig M (2012) Ml confidential: machine learning on encrypted data. In: International conference on information security and cryptology. Springer, pp 1–21

Gustafson DE, Kessel WC (1978) Fuzzy clustering with a fuzzy covariance matrix. In: IEEE conference on decision and control including the 17th symposium on adaptive processes, vol 1979. IEEE, pp 761–766

Hesamifard E, Takabi H, Ghasemi M (2017) CryptoDL: deep neural networks over encrypted data. arXiv:1711.05189

Hitaj B, Ateniese G, Perez-Cruz F (2017) Deep models under the GAN: information leakage from collaborative deep learning. In: Proceedings of the 2017 ACM SIGSAC conference on computer and communications security, pp 603–618

Hubara I, Courbariaux M, Soudry D, El-Yaniv R, Bengio Y (2017) Quantized neural networks: training neural networks with low precision weights and activations. J Mach Learn Res 18(1):6869–6898

Hunt T, Zhu Z, Xu Y, Peter S, Witchel E (2018) Ryoan: a distributed sandbox for untrusted computation on secret data. ACM Trans Comput Syst (TOCS) 35(4):1–32

Hunt T, Song C, Shokri R, Shmatikov V, Witchel E (2018) Chiron: privacy-preserving machine learning as a service. arXiv:1803.05961

Ioffe S, Szegedy C (2015) Batch normalization: accelerating deep network training by reducing internal covariate shift. arXiv:1502.03167

Jiang X, Kim M, Lauter K, Song Y (2018) Secure outsourced matrix computation and application to neural networks. In: Proceedings of the 2018 ACM SIGSAC conference on computer and communications security, pp 1209–1222

Juvekar C, Vaikuntanathan V, Chandrakasan A (2018) {GAZELLE}: a low latency framework for secure neural network inference. In: 27th {USENIX} security symposium ({USENIX} security 18), pp 1651–1669

Keller M (2020) MP-SPDZ: a versatile framework for multi-party computation. In: Proceedings of the 2020 ACM SIGSAC conference on computer and communications security, pp 1575–1590

Kim S-J, Magnani A, Boyd S (2006) Robust fisher discriminant analysis. In: Advances in neural information processing systems, pp 659–666

Kim M, Smaragdis P (2016) Bitwise neural networks. arXiv:1601.06071

Krishnapuram R, Keller JM (1996) The possibilistic c-means algorithm: insights and recommendations. IEEE Trans Fuzzy Syst 4(3):385–393

Kumar N, Rathee M, Chandran N, Gupta D, Rastogi A, Sharma R (2019) CrypTFlow: secure tensorflow inference. arXiv:1909.07814

Law A, Leung C, Poddar R, Popa RA, Shi C, Sima O, Yu C, Zhang X, Zheng W (2020) Secure collaborative training and inference for xgboost. In: Proceedings of the 2020 workshop on privacy-preserving machine learning in practice, pp 21–26

Liu J, Tian Y, Zhou Y, Xiao Y, Ansari N (2020) Privacy preserving distributed data mining based on secure multi-party computation. Comput Commun 153:208–216

Liu J, Juuti M, Lu Y, Asokan N (2017) Oblivious neural network predictions via minionn transformations. In: Proceedings of the 2017 ACM SIGSAC conference on computer and communications security, pp 619–631

Liu W, Pan F, Wang XA, Cao Y, Tang D (2018) Privacy-preserving all convolutional net based on homomorphic encryption. In: International conference on network-based information systems. Springer, pp 752–762

McKeen F, Alexandrovich I, Anati I, Caspi D, Johnson S, Leslie-Hurd R, Rozas C (2016) Intel® software guard extensions (intel® SGX) support for dynamic memory management inside an enclave. Proc Hardw Arch Supp Securand Privacy 2016:1–9

Mohassel P, Rindal P (2018) ABY3: a mixed protocol framework for machine learning. In: Proceedings of the 2018 ACM SIGSAC conference on computer and communications security, pp 35–52

Mohassel P, Zhang Y (2017) Secureml: a system for scalable privacy-preserving machine learning. In: IEEE symposium on security and privacy (SP). IEEE, pp 19–38

Ohrimenko O, Schuster F, Fournet C, Mehta A, Nowozin S, Vaswani K, Costa M (2016) Oblivious multi-party machine learning on trusted processors. In: 25th USENIX security symposium (USENIX security 16), pp 619–636

Papernot N, Abadi M, Erlingsson U, Goodfellow I, Talwar K (2016) Semi-supervised knowledge transfer for deep learning from private training data. arXiv:1610.05755

Park S, Byun J, Lee J, Cheon JH, Lee J (2020) He-friendly algorithm for privacy-preserving SVM training. IEEE Access, vol 8, pp 57 414–57 425

Park J, Kim DS, Lim H (2020) Privacy-preserving reinforcement learning using homomorphic encryption in cloud computing infrastructures. IEEE Access, vol 8, pp 203 564–203 579

Paszke A, Gross S, Massa F, Lerer A, Bradbury J, Chanan G, Killeen T, Lin Z, Gimelshein N, Antiga L et al (2019) PyTorch: an imperative style, high-performance deep learning library. In: Advances in neural information processing systems, pp 8024–8035

Ramírez DH, Auñón J (2020) Privacy preserving k-means clustering: a secure multi-party computation approach. arXiv:2009.10453

Riazi MS, Weinert C, Tkachenko O, Songhori EM, Schneider T, Koushanfar F (2018) Chameleon: a hybrid secure computation framework for machine learning applications. In: Proceedings of the 2018 on Asia conference on computer and communications security, pp 707–721

Rouhani BD, Riazi MS, Koushanfar F (2018) DeepSecure: scalable provably-secure deep learning. In: Proceedings of the 55th annual design automation conference, pp 1–6

Ryffel T, Trask A, Dahl M, Wagner B, Mancuso J, Rueckert D, Passerat-Palmbach J (2018) A generic framework for privacy preserving deep learning. arXiv:1811.04017

Sanyal A, Kusner MJ, Gascon A, Kanade V (2018) TAPAS: tricks to accelerate (encrypted) prediction as a service. arXiv:1806.03461

Schwarzenberg-Czerny A (1995) On matrix factorization and efficient least squares solution. Astron Astrophys Suppl Ser 110:405

Shokri R, Shmatikov V (2015) Privacy-preserving deep learning. In: Proceedings of the 22nd ACM SIGSAC conference on computer and communications security, pp 1310–1321

Shokri R, Stronati M, Song C, Shmatikov V (2017) Membership inference attacks against machine learning models. In: IEEE symposium on security and privacy (SP). IEEE, pp 3–18

Skurichina M, Duin RP (2002) Bagging, boosting and the random subspace method for linear classifiers. Pattern Anal Appl 5(2):121–135

Smyth P (2000) Model selection for probabilistic clustering using cross-validated likelihood. Stat Comput 10(1):63–72

So J, Guler B, Avestimehr AS, Mohassel P (2019) CodedPrivateML: a fast and privacy-preserving framework for distributed machine learning. arXiv:1902.00641

Tramer F, Boneh D (2018) SLALOM: fast, verifiable and private execution of neural networks in trusted hardware. arXiv:1806.03287

Tramèr F, Zhang F, Juels A, Reiter MK, Ristenpart T (2016) Stealing machine learning models via prediction APIS. In: 25th {USENIX} security symposium ({USENIX} security 16), pp 601–618

Tran A-T, Luong T-D, Karnjana J, Huynh V-N (2021) An efficient approach for privacy preserving decentralized deep learning models based on secure multi-party computation. Neurocomputing 422:245–262

Van Bulck J, Minkin M, Weisse O, Genkin D, Kasikci B, Piessens F, Silberstein M, Wenisch TF, Yarom Y, Strackx R (2018) Foreshadow: extracting the keys to the intel {SGX} kingdom with transient out-of-order execution. In: 27th {USENIX} security symposium ({USENIX} security 18), pp 991–1008

Wagh S, Gupta D (2019) SecureNN: 3-party secure computation for neural network training. Proc Priv Enhanc Technol 3:26–49

Xue H, Huang Z, Lian R, Qiu W, Guo J, Wang S, Gong Z (2018) Distributed large scale privacy-preserving deep mining. In: 2018 IEEE third international conference on data science in cyberspace (DSC). IEEE, pp 418–422

Xu R, Joshi JB, Li C (2019) Cryptonn: training neural networks over encrypted data. In: 2019 IEEE 39th international conference on distributed computing systems (ICDCS). IEEE, pp 1199–1209

Yao AC-C (1986) How to generate and exchange secrets. In: 27th annual symposium on foundations of computer science, vol 1986. IEEE, pp 162–167

Zhang Q, Yang LT, Castiglione A, Chen Z, Li P (2019) Secure weighted possibilistic c-means algorithm on cloud for clustering big data. Inf Sci 479:515–525

Chapter 4
Pros and Cons of X-Based PPDL

Abstract This chapter discusses the comparison of all of privacy-preserving deep learning methods, highlighting the pros and cons of each method based on privacy parameters, used specific neural network and dataset type from the point of performance. We also provide our analysis about the weakness of each privacy-preserving deep learning method and our feasible solution to address their weakness.

4.1 Metrics for Comparison

To compare the performances of each surveyed article, we used two kinds of metrics, qualitative metrics and quantitative metrics. Figure 4.1 shows the metrics for the surveyed PPDL works in this chapter. Qualitative metrics include Privacy of Client (PoC), Privacy of Model (PoM), and Privacy of Result (PoR). PoC means that neither the model owner nor the cloud server or any other party knows about the client data. PoM means that neither the client nor the cloud server or any other party knows about the DL model. PoR means that neither the model owner nor the cloud server or any other party can obtain the information about the prediction result. Quantitative metrics include accuracy and inference time. Accuracy means the percentage of correct predictions made by a PPDL model. The inference time is the time needed by the model to perform encryption/decryption, send data from the client to the server, and execute the classification process. We measured the average accuracy and inference time of each method. Then, we set the average value as the relative evaluation. If the accuracy value is higher than average, the accuracy of the proposed method is good. Furthermore, if the run time and data transfer are lower than average, the run time and data transfer of the proposed methods are good. We used the comparison data from the respective papers as we believe they are the best result to be achieved. We did not re-execute their codes as not all of the codes are open to the public. We focused the comparison on the Hybrid PPDL method, which combines classical PP with various deep learning practices.

K. Kim and H. C. Tanuwidjaja, *Privacy-Preserving Deep Learning*,
SpringerBriefs on Cyber Security Systems and Networks,
https://doi.org/10.1007/978-981-16-3764-3_4

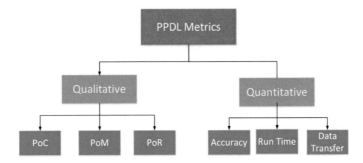

Fig. 4.1 Metrics for surveyed PPDL works

4.2   Comparison of X-Based PPDL

We divided our comparison table into two types: performance comparison I and performance comparison II in Tables 4.1 and 4.2, respectively. To compare the performance of each surveyed paper, we used the privacy metrics and performance metrics defined before. The privacy metrics include Privacy of Client (PoC), Privacy of Model (PoM), and Privacy of Result (PoR). The performance metrics include accuracy, run time, and data transfer.

4.3   Weaknesses and Possible Solutions of X-Based PPDL

In this section, we will discuss the challenges and weaknesses of utilizing PPDL for MLaaS from the papers that we surveyed. To analyze the limitations, we divided the PPDL approach into two main categories based on the type of transmission: the model parameter transmission approach and the data transmission approach. The model parameter transmission approach means that the model parameter is transmitted from the client to the server while the local data is kept by the client, and the training is performed on the client side. On the other hand, the data transmission approach means that the client data itself is transferred to the server for the training process. In short, the challenges and weaknesses of state-of-the-art PPDL methods are shown in Fig. 4.2.

4.3.1   Model Parameter Transmission Approach

In this approach, during the training process, a model parameter is transmitted instead of the training data. PPDLs based on distributed machine learning and federated learning are included in this scenario. In distributed learning (Dean et al. 2012;

Table 4.1 Performance comparison I

References	Used PP technique					Privacy parameters			ML/DL method	Dataset type
	HE	Secure-MPC	Oblivious transfer	Differential privacy	Secure enclaves	PoC	PoM	PoR		
CryptoNets (2016)	O	X	X	X	X	O	X	O	CNN	Image
PATE (2016)	X	X	X	O	X	O	O	X	GAN	Image
GAZELLE (2018)	O	O	O	X	X	O	O	O	CNN	Image
TAPAS (2018)	O	X	X	X	X	O	O	O	CNN	Image
FHE DiNN (2018)	O	X	X	X	X	O	O	O	CNN	Image
E2DM (2018)	O	X	X	X	X	O	O	O	CNN	Image
ABY3 (2018)	X	O	O	X	X	X	O	O	CNN	Image
Xue18 (2018)	O	X	X	X	X	O	X	O	CNN	Image
Aono17 (2017)	O	X	X	X	X	O	X	O	NN	Image
Zhang19 (2019)	O	X	X	X	X	O	X	O	C-Means algorithm	Image
ML Confidential (2012)	O	X	X	X	X	O	X	O	Linear Means Classifier	CSV
Chabanne17 (2017)	O	X	X	X	X	O	X	O	DNN	Image
CryptoDL (2017)	O	X	X	X	X	O	X	O	CNN	Image
Liu18 (2018)	O	X	X	X	X	O	X	O	CNN	Image
SecureML (2017)	X	O	O	X	X	X	O	O	NN	Image
DeepSecure (2018)	X	O	O	X	X	X	O	O	CNN	Image
MiniONN (2017)	X	O	O	X	X	X	O	O	CNN	Image
SLALOM (2018)	X	X	X	X	O	O	O	O	DNN	Image
SecureNN (2019)	X	O	O	X	X	X	O	O	CNN	Image
Ryffel18 (2018)	X	O	O	O	X	O	O	O	Federated learning	CSV

(continued)

Table 4.1 (continued)

References	Used PP technique						Privacy parameters				ML/DL method	Dataset type
	HE	Secure-MPC	Oblivious transfer	Differential privacy	Secure enclaves		PoC	PoM	PoR			
Faster Crytonets (2018)	O	X	X	X	X		O	X	O		CNN	Image
CodedPrivateML (2019)	X	O	O	X	X		O	O	O		Logistic Regression	Image
Chiron (2018)	X	X	X	X	O		O	O	O		DNN	Image
Bu19 (2019)	X	X	X	O	X		O	O	O		DNN	Image
CrypTFlow (2019)	X	O	O	X	O		O	O	O		DNN	Image
Chameleon (2018)	X	O	O	X	X		X	O	O		CNN	Image
Chase17 (2017)	X	O	O	O	X		X	O	O		CNN	Image
Ohrimenko16 (2016)	X	O	O	X	O		O	X	O		DNN	CSV

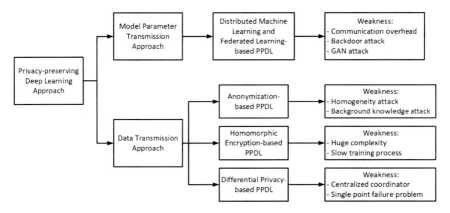

Fig. 4.2 The challenges and weaknesses of state-of-the-art PPDL methods

Hamm et al. 2015; So et al. 2019; McMahan et al. 2016), data owners keep their own data secret without revealing this data to another party. During each training stage, participants send their locally computed parameters to the server. By doing this, the participants can learn collaboratively without revealing their own data (Song and Chai 2018). On the other hand, in federated learning (Truex et al. 2019; Hardy et al. 2017; Mugunthan et al. 2020), model provider sends the model to participants. Then, each participant executes the training process using their local data, resulting in an updated model. After that, the updated model is sent back to the model provider. The model provider will measure the average value of the gradient descent and update the model. We can see that the model parameter transmission approach reduces the communication overhead but increases the local computation overhead. This learning approach is vulnerable to backdoor attacks (Sun et al. 2019) and GAN-based attacks (Wang et al. 2019).

4.3.2 Data Transmission Approach

In this approach, the participants send their data to the training server. Some PPDL methods that belong to this class are anonymization, HE, and DP-based PPDL. The main purpose of the anonymization technique is to remove the correlation between the data owner and the data entries. However, it requires a trusted coordinator to perform the anonymization process and distribute the result to the participants. It is also vulnerable to a single point of failure as a trusted proxy needs to perform the anonymization process and send the result to the participants (Vergara-Laurens et al. 2016). HE-based PPDL does not require key distribution management because the computation can be performed on encrypted data. However, it has limitations in the computation format. The computation is limited to a polynomial of bounded degree; thus, it works in a linear nature. Another weakness of HE-based PPDL is the slow training process as it has huge complexity, and the computation process will lead to data swelling (Jiang et al. 2019).

A bootstrapping idea (Ducas and Micciancio 2015; Chillotti et al. 2016; Cheon et al. 2018) has been introduced to solve this problem by reducing the complexity and the computation time. The majority of the work focuses on polynomial approximation for non-linear operations. The main goal of DP-based PPDL is to perturb the sample data for the training process. It is often used for data such as histograms or tables. The main weakness of DP-based PPDL is its centralized nature. One main trusted coordinator that is responsible for data collection and giving the response to queries from participants. This trusted coordinator is vulnerable to a single point of failure. If this kind of failure occurs and each participant perturbs the training data, the model will yield poor accuracy (Jiang et al. 2019). Thus, a centralized coordinator is very susceptible to the single point of failure problem. In a nutshell, we can conclude that the data transmission approach reduces the computation overhead but increase the communication overhead.

4.3.3 Analysis and Summary

After discussing the challenges and weaknesses in PPDL from the two categories above, we summarize the two main problems in PPDL: the computation overhead and communication overhead.

4.3.3.1 Computation Overhead

One of the most important issues in PPDL is the computation overhead. The overhead issues occur during the HE process, deep learning training (including inferencing), and data perturbation. The amount of overhead is related to the total number of gradients. Many researches try to address this issue by proposing bootstrapping methods. Currently, utilizing PPDL for large-scale service is not feasible in real life because of this scalability problem. This leads to the introduction of privacy-preserving federated learning which is designed for collaborative learning among many participants.

4.3.3.2 Communication Overhead

In PPDL, communication overhead occurs during the interaction among clients, model providers, and server providers. In particular, we can categorize communication overhead into the HE process, additive or multiplicative perturbation, and iterative communication. Performing multiple epochs of gradient descent will also caused a communication delay since the number of gradient evaluation increases too. In the distributed machine learning scenario, including the federated learning,

Table 4.2 Performance comparison II

PPDL type	Proposed scheme	Accuracy (%)	Inference time (s)
HE-based PPDL	Cryptonets (2016)	Good (98.95)	Bad (297.5)
	Aono17 (2017)	Good (97.00)	Good (120)
	Chabanne17 (2017)	Good (99.30)	–
	CryptoDL (2017)	Good (99.52)	Bad (320)
	TAPAS (2018)	Good (98.60)	Good (147)
	FHE-DiNN (2018)	Bad (96.35)	Good (1.64)
	E2DM (2018)	Good (98.10)	Good (28.59)
	Xue18 (2018)	Good (99.73)	–
	Liu18 (2018)	Good (98.97)	Bad (477.6)
	Faster cryptonets (2018)	Bad (80.61)	–
	CryptoNN (2019)	Bad (95.48)	–
	Zhang19 (2019)	–	–
	ML confidential (2012	Bad (95.00)	Bad (255.7)
Secure MPC-based PPDL	SecureML (2017)	Bad (93.40)	Good (4.88)
	MiniONN (2017)	Good (98.95)	Good (9.32)
	ABY3 (2018)	Bad (94.00)	Good (0.01)
	DeepSecure (2018)	Good (98.95)	Good (9.67)
	Chameleon (2018)	Good (99.00)	Good (2.24)
	SecureNN (2019)	Good (99.15)	Good (0.076)
	CodedPrivateML (2019)	Bad (95.04)	Bad (110.9)
Differential privacy-based PPDL	PATE (2016)	Good (98.10)	–
	Bu19 (2019)	Good (98.00)	–
Secure enclaves-based PPDL	Chiron (2018	Bad (89.56)	–
	SLALOM (2018	Bad (92.4)	–
Hybrid-based PPDL	Ohrimenko16 (2016)	Good (98.7)	Good (2.99)
	Chase17 (2017)	Good (98.9)	–
	Gazelle (2018)	–	Good (0.03)
	Ryffel18 (2018)	Bad (70.3)	–
	CrypTFlow (2019)	Bad (93.23)	Good (30)

this factor is the main scalability problem that becomes the main issue. The iterative communications to exchange data and model parameters between each party will produce a significant overhead problem.

References

Aono Y, Hayashi T, Wang L, Moriai S et al (2017) Privacy-preserving deep learning via additively homomorphic encryption. IEEE Trans Inf Forensics Secur 13(5):1333–1345

Bourse F, Minelli M, Minihold M, Paillier P (2018) Fast homomorphic evaluation of deep discretized neural networks. In: Annual international cryptology conference. Springer, pp 483–512

Bu Z, Dong J, Long Q, Su WJ (2019) Deep learning with gaussian differential privacy. arXiv:1911.11607

Chabanne H, de Wargny A, Milgram J, Morel C, Prouff E (2017) Privacy-preserving classification on deep neural network. IACR Cryptol ePrint Arch 2017:35

Chase M, Gilad-Bachrach R, Laine K, Lauter KE, Rindal P (2017) Private collaborative neural network learning. IACR Cryptol ePrint Arch 2017:762

Cheon JH, Han K, Kim A, Kim M, Song Y (2018) Bootstrapping for approximate homomorphic encryption. In: Annual international conference on the theory and applications of cryptographic techniques. Springer, pp 360–384

Chillotti I, Gama N, Georgieva M, Izabachene M (2016) Faster fully homomorphic encryption: bootstrapping in less than 0.1 s. In: International conference on the theory and application of cryptology and information security. Springer, pp 3–33

Chou E, Beal J, Levy D, Yeung S, Haque A, Fei-Fei L (2018) Faster cryptonets: leveraging sparsity for real-world encrypted inference. arXiv:1811.09953

Dean J, Corrado G, Monga R, Chen K, Devin M, Mao M, Ranzato M, Senior A, Tucker P, Yang K et al (2012) Large scale distributed deep networks. In: Advances in neural information processing systems, pp 1223–1231

Ducas L, Micciancio D (2015) FHEW: bootstrapping homomorphic encryption in less than a second. In: Annual international conference on the theory and applications of cryptographic techniques. Springer, pp 617–640

Gilad-Bachrach R, Dowlin N, Laine K, Lauter K, Naehrig M, Wernsing J (2016) Cryptonets: applying neural networks to encrypted data with high throughput and accuracy. In: International conference on machine learning, pp 201–210

Graepel T, Lauter K, Naehrig M (2012) Ml confidential: machine learning on encrypted data. In: International conference on information security and cryptology. Springer, pp 1–21

Hamm J, Champion AC, Chen G, Belkin M, Xuan D (2015) Crowd-ml: a privacy-preserving learning framework for a crowd of smart devices. In: 2015 IEEE 35th international conference on distributed computing systems. IEEE, pp 11–20

Hardy S, Henecka W, Ivey-Law H, Nock R, Patrini G, Smith G, Thorne B (2017) Private federated learning on vertically partitioned data via entity resolution and additively homomorphic encryption. arXiv:1711.10677

Hesamifard E, Takabi H, Ghasemi M (2017) Cryptodl: deep neural networks over encrypted data. arXiv:1711.05189

Hunt T, Song C, Shokri R, Shmatikov V, Witchel E (2018) Chiron: privacy-preserving machine learning as a service. arXiv:1803.05961

Jiang X, Kim M, Lauter K, Song Y (2018) Secure outsourced matrix computation and application to neural networks. In: Proceedings of the 2018 ACM SIGSAC conference on computer and communications security, pp 1209–1222

Jiang L, Tan R, Lou X, Lin G (2019) On lightweight privacy-preserving collaborative learning for internet-of-things objects. In: Proceedings of the international conference on internet of things design and implementation, pp 70–81

Juvekar C, Vaikuntanathan V, Chandrakasan A (2018) *GAZELLE*: a low latency framework for secure neural network inference. In: 27th *USENIX* security symposium (*USENIX*) security 18), pp 1651–1669

Kumar N, Rathee M, Chandran N, Gupta D, Rastogi A, Sharma R (2019) Cryptflow: secure tensorflow inference. arXiv:1909.07814

Liu J, Juuti M, Lu Y, Asokan N (2017) Oblivious neural network predictions via minionn transformations. In: Proceedings of the 2017 ACM SIGSAC conference on computer and communications security, pp 619–631

Liu W, Pan F, Wang XA, Cao Y, Tang D (2018) Privacy-preserving all convolutional net based on homomorphic encryption. In: International conference on network-based information systems. Springer, pp 752–762

McMahan HB, Moore E, Ramage D, Hampson S et al (2016) Communication-efficient learning of deep networks from decentralized data. arXiv:1602.05629

Mohassel P, Rindal P (2018) Aby3: a mixed protocol framework for machine learning. In: Proceedings of the 2018 ACM SIGSAC conference on computer and communications security, pp 35–52

Mohassel P, Zhang Y (2017) Secureml: a system for scalable privacy-preserving machine learning. In: IEEE symposium on security and privacy (SP). IEEE, 19–38

Mugunthan V, Peraire-Bueno A, Kagal L (2020) Privacyfl: a simulator for privacy-preserving and secure federated learning. arXiv:2002.08423

Ohrimenko O, Schuster F, Fournet C, Mehta A, Nowozin S, Vaswani K, Costa M (2016) Oblivious multi-party machine learning on trusted processors. 25th *USENIX* security symposium

Papernot N, Abadi M, Erlingsson U, Goodfellow I, Talwar K (2016) Semi-supervised knowledge transfer for deep learning from private training data. arXiv:1610.05755

Riazi MS, Weinert C, Tkachenko O, Songhori EM, Schneider T, Koushanfar F (2018) Chameleon: a hybrid secure computation framework for machine learning applications. In: Proceedings of the 2018 on Asia conference on computer and communications security, pp 707–721

Rouhani BD, Riazi MS, Koushanfar F (2018) Deepsecure: scalable provably-secure deep learning. In: Proceedings of the 55th annual design automation conference, pp 1–6

Ryffel T, Trask A, Dahl M, Wagner B, Mancuso J, Rueckert D, Passerat-Palmbach J (2018) A generic framework for privacy preserving deep learning. arXiv:1811.04017

Sanyal A, Kusner MJ, Gascon A, Kanade V (2018) Tapas: tricks to accelerate (encrypted) prediction as a service. arXiv:1806.03461

So J, Guler B, Avestimehr AS, Mohassel P (2019) Codedprivateml: a fast and privacy-preserving framework for distributed machine learning. arXiv:1902.00641

Song G, Chai W (2018) Collaborative learning for deep neural networks. In: Advances in neural information processing systems, pp 1832–1841

Sun Z, Kairouz P, Suresh AT, McMahan HB (2019) Can you really backdoor federated learning? arXiv:1911.07963

Tramer F, Boneh D (2018) Slalom: fast, verifiable and private execution of neural networks in trusted hardware. arXiv:1806.03287

Truex S, Baracaldo N, Anwar A, Steinke T, Ludwig H, Zhang R, Zhou Y (2019) A hybrid approach to privacy-preserving federated learning. In: Proceedings of the 12th ACM workshop on artificial intelligence and security, pp 1–11

Vergara-Laurens IJ, Jaimes LG, Labrador MA (2016) Privacy-preserving mechanisms for crowd-sensing: survey and research challenges. IEEE Internet Things J 4(4):855–869

Wagh S, Gupta D, Chandran N (2019) Securenn: 3-party secure computation for neural network training. Proc Priv Enhanc Technol 3:26–49

Wang Z, Song M, Zhang Z, Song Y, Wang Q, Qi H (2019) Beyond inferring class representatives: User-level privacy leakage from federated learning. In: IEEE INFOCOM 2019-IEEE conference on computer communications. IEEE, pp 2512–2520

Xue H, Huang Z, Lian H, Qiu W, Guo J, Wang S, Gong Z (2018) Distributed large scale privacy-preserving deep mining. In: 2018 IEEE third international conference on data science in cyberspace (DSC). IEEE, pp 418–422

Xu R, Joshi JB, Li C (2019) Cryptonn: training neural networks over encrypted data. In: 2019 IEEE 39th international conference on distributed computing systems (ICDCS). IEEE, pp 1199–1209

Zhang Q, Yang LT, Castiglione A, Chen Z, Li P (2019) Secure weighted possibilistic c-means algorithm on cloud for clustering big data. Inf Sci 479:515–525

Chapter 5
Privacy-Preserving Federated Learning

Abstract In this chapter, we introduce the emerging application of privacy-preserving federated learning in a coordinated way among multi-party. We suggest the function specific Privacy-Preserving Federated Learning (PPFL) to provide fairness,integrity, correctness, adaptiveness and flexibility. Application specific PPFL includes mobile devices, medical imaging, traffic flow prediction and healthcare, Android malware detection, and edge computing.

5.1 Overview

Privacy-Preserving Federated Learning (PPFL) is a utilization of federated learning in privacy-preserved environment. Federated learning is a machine learning framework that enables collaborative learning between many parties without exchanging data samples among the participants. PPFL aims to protect the privacy of participants using the well known privacy-preserving method, such as HE, secure MPC, DP, and secure enclaves. Figure 5.1 shows the Architecture of PPFL. A collaborator is the trusted authority who coordinate the collaborative learning. First, each party executes local training and sends encrypted gradient to the the collaborator. The collaborator aggregates all gradients in the server, then sends back the updated model to each party. Then, each party updates its own model and do local training process again. The process repeats until the optimum model has been achieved.

As a distributed system, PPFL is different from the traditional centralized machine learning system. It has many contributing clients so that PPFL needs to be scalable to large number of participants (McMahan et al. 2017). There are also various kinds of clients; some have a few samples, while the others have a big number of data. Among participants, the data distribution can be different since it is random. A latency issue also appears quite often since the Internet speed between each participant is not the same. Some participant can be mobile and cause an unstable communication. So, achieving lower latency is an important issue in PPFL.

K. Kim and H. C. Tanuwidjaja, *Privacy-Preserving Deep Learning*,
SpringerBriefs on Cyber Security Systems and Networks,
https://doi.org/10.1007/978-981-16-3764-3_5

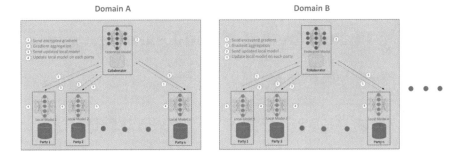

Fig. 5.1 Architecture of PPFL

5.2   Function Specific PPFL

Figure 5.2 shows the Classification of function specific PPFL.

5.2.1   Fairness

FPPDL (Lyu et al. 2020) is a FL-based PPDL framework with fairness consideration. It combines DP method using GAN and a three-layer encryption scheme to guarantee the privacy of all parties. FPPDL emphasizes fairness factor, which means that each party get shared model based on its contribution. By doing this, the unfairness issue that all parties are given the same model without regard to their contribution, has been solved. The authors design a "earn and pay" system, where party with high contributions are awarded with a better performing local model than the parties with lower contribution. It will encourage participants to share their gradients actively.

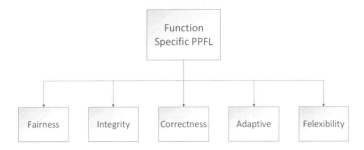

Fig. 5.2 Classification of function specific PPFL

5.2.2 Integrity

Chen20 (2020) proposed a PPFL scheme that guarantees the integrity of all parties. Integrity means that no participant attempts to do any dishonest action, for example delaying the training process or tampering the training model. The scheme utilizes trusted execution environment to ensure that all participants execute the learning algorithm in a correct way. As a result, if there is any attempt to break the availability of the model, it will be automatically detected. The experiments showed that the proposed scheme has successfully ensured the integrity of all participants, while keeping their privacy preserved.

5.2.3 Correctness

Zhang20 (2020) proposed a verifiable PPFL framework with low communication and computation cost. Verifiable PPDL means that the correctness of aggregated gradient during the training process can be guaranteed. If the aggregation server falsify the aggregated gradient value, it will be detected as malicious attempt. Therefore, as long as all participants follow the system rule, the aggregated gradient value should be correct. The framework also guarantees the data privacy of each participant since the aggregation server cannot obtain the original data from the aggregated value. The authors successfully proved that the proposed method still has high accuracy and efficiency even though a verification function was added.

5.2.4 Adaptive

Adaptive Privacy-Preserving Federated Learning (APFL) (Liu et al. 2020) is a PP framework that combines DP with federated learning. DP is much more efficient than HE and Secure MPC. However, due to its noise; it always has trade-off between accuracy and privacy. APFL tries to address this issue and seek for the optimal balance. They calculate the contribution of each class to the output by using a propagation algorithm. APFL significantly reduce the noise in the final result by injecting adaptive noise. The adaptive noise means that the noise injection will be based on the importance of each data. In order to improve the accuracy of the model, a Randomized Privacy-Preserving Adjustment Technology (RPAT) is added to the framework. RPAT enables users to personalize parameter to filter the extra noise. The experiments showed that APFL achieved high accuracy and efficiency compared to the existing framework.

5.2.5   Flexibility

Poseidon (Sav et al. 2020) is a privacy-preserving neural network training in a federated environment. It utilizes multi-party lattice-based cryptography to ensure the privacy of the data and model. Poseidon supports single instruction multiple data operation on encrypted data to do efficient back propagation process. It also tries to address the cost issue by leveraging arbitrary linear transformation within the bootstrapping operation. The pooling process or large number of operations are the main target of the boostrapping. The experiment showed that Poseidon has successfully trains a three layer neural network using MNIST dataset with 60,000 instances and 784 features in less than 2 h.

5.3   Application Specific PPFL

Figure 5.3 shows the Classification of application specific PPFL.

5.3.1   Mobile Devices

Bonawitz17 (2017) designed a secure aggregation protocol for PPFL in mobile devices. The proposed protocol works on high dimensional vectors and robust to any kind of user. It can handle the case when a user suddenly drops out during a training process. It is also highly efficient in the term of communication with security in unauthenticated network model. Furthermore, Bonawitz19 (2019) enhanced their work and built a scalable PPFL framework for mobile devices using tensorflow. The framework addresses some issues, such as: unreliable connectivity, interrupted execution, limited device storage, and computation resources.

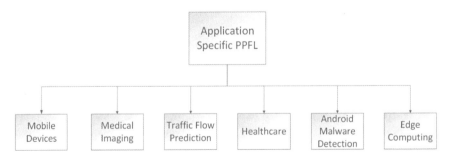

Fig. 5.3 Classification of application specific PPFL

5.3.2   Medical Imaging

Kaissis20 (2020) proposed a PPFL study in medical imaging. They want to address the legal and ethical issue of patients' privacy in the application of artificial intelligence for medical imaging. There is no standardized electronic medical records yet, that is why a strict requirement must be defined. The proposed framework prevents patient privacy compromise while doing AI learning with patients' medical imaging. They analyze the combination of federated learning with HE, Secure MPC, and DP.

5.3.3   Traffic Flow Prediction

Federated Learning-based Gated Recurrent Unit Neural Network Algorithm (Fed-GRU) (Liu et al. 2020) is a PPFL method for traffic flow prediction. It is different from other centralized learning method by implementing secure parameter aggregation mechanism. The federated learning algorithm enables FedGRU to avoid direct raw data sharing between participants. In order to improve the scalability, they use a random sub sampling protocol. By doing this, the communication overhead can be reduced, which is suitable for large-scale system. They also develop a clustering method to integrate the optimal global model and spatial traffic flow data to improve the prediction accuracy.

5.3.4   Healthcare

Grama20 (2020) proposed a robust aggregation method in PPFL using healthcare dataset. They did experiments to evaluate the impact of DP and k-anonymity to the model accuracy. Typical PPFL methods are sensitive to local model updates. It can be a hindrance to make a robust global model. Their experiments showed that the proposed method successfully detected and discarded malicious local parties during the training process. They also showed that DP did not significantly affect the time consumed during the aggregation process.

5.3.5   Android Malware Detection

Hsu20 (2020) proposed a PPFL system for Android malware detection. The system uses static analysis and meta-data analysis; combined with federated learning to preserve participants' privacy. The system design is based on edge computing to reduce latency and communication cost. They combines SVM with Secure MPC in a PPFL system using Android malware dataset. The experiments result showed that the proposed system has relatively similar accuracy with the same amount of data compared to traditional machine learning system with no privacy preservation.

5.3.6   Edge Computing

Privacy-preserving Asynchronous Federated Learning Mechanism for Edge Network Computing (PAFLM) (Lu et al. 2020) is a PPFL method on edge computing that allows multiple edge nodes to achieve privacy-preserved data training. Edge computing offers a faster cloud system by reducing the delay of data transmission. It can be done by doing data processing on the edge node. The proposed method compresses the communication between participants and server. The authors design a gradient compression algorithm to reduce the communication and lower the possibility of privacy breach. It also utilizes asynchronous federated learning that allows a node to join or quit anytime, which is suitable for highly mobile devices.

5.4   Summary

Table 5.1 shows the Key features of PPFL methods we have discussed here.

In a nutshell, there are two main properties of PPFL: concurrency and model consistency (Briggs et al. 2020). Concurrency means that the ability to do parallel calculation across multiple node in the same time. Each node has to communicate regularly to other nodes when computing the activation of its neurons. However, there will be a bottleneck in the slowest computing node. In order to resolve this issue, Dean12 (2012) distributes one or more layers on each node; ensuring that only working nodes communicate with each other. Even though this method solves the excessive communication issue, it still requires copying mini-batch data to all nodes. A better parallelism technique is proposed by Ben19 (2019) that partitions the training data and copies each partition to all computing nodes. The weight updates are reduced on each iteration using message parsing interface. This method solves the scalability issue since it is able to handle a large amount of data. Model consistency means that all participants received the same model during the training process. The model should contain the same parameter values. The parameter is updated to global server during each training iteration (Li et al. 2014). Then, the global server runs the aggregation process to synchronize the distributed model.

PPFL is vulnerable to GAN-based attack. It was introduced by Wang19 (2019). Firstly, GAN generates samples that looks similar to the training data. Since it do not have access to the original training data, the fake samples are produced by interacting with the discriminative model. The goal of this attack is deceiving the discriminative model into believing that the produced sample is real. While the goal of the discriminative model is to detect the synthesis sample. The experiment showed that

Table 5.1 Key features of PPFL methods

References	Year	Key features	Functionality	Target application
Bonawitz17 (2017)	2017	Utilizing gaussian-based DP for PPFL with reduced loss	–	Mobile devices
Bonawitz19 (2019)	2019	Proposing secure MPC-based PPFL with optional secure aggregation protocol	–	Mobile devices
Kaissis20 (2020)	2020	Analyzing the combination of FL with HE, Secure MPC, and DP	–	Medical imaging
FedGRU (Liu et al. 2020)	2020	Developing a clustering method to integrate the optimal global model	–	Traffic flow prediction
Chen20 (2020)	2020	Proposing trusted execution environment for PPFL to detect any dishonest action	Integrity	–
FPPDL (Lyu et al. 2020)	2020	Utilizing GAN for PPFL with three layer encryption scheme	Fairness	–
APFL (Liu et al. 2020)	2020	Combining differential privacy with federated learning with noise filtering	Adaptive	–
Poseidon (Sav et al. 2020)	2020	Supporting single instruction multiple data operation to do efficient backpropagation process	Flexibility	–
Grama20 (2020)	2020	Evaluating the impact of differential privacy and k-anonymity to the PPFL model accuracy	–	Healthcare
PAFLM (Lu et al. 2020)	2020	Designing a gradient compression algorithm to reduce the communication cost	–	Edge computing
Hsu20 (2020)	2020	Utilizing static analysis and meta-data analysis with federated learning to preserve participants' privacy	–	Android malware detection
Zhang20 (2020)	2020	Proposing PPFL framework with low communication and computation cost	Correctness	–

their proposed attack successfully exposed the PPFL model, proving it has security vulnerability to GAN attack.

From all papers that we have surveyed, we can summarize that there are four main challenges in implementing PPDL in IoT. The first challenge is finding the optimum hyper-parameter for the model trained on. In order to get a good model, a research in efficient training need to be conducted. The second challenge is time cost. Since federated learning is an expensive and time-consuming method, the cost will grow over time. The third challenge is finding a better PP method. The current privacy-preserving method has its own issue. For example, homomorphic encryption with complexity issue, secure MPC with adversarial issue, and differential privacy with trade-off issue. The last challenge is about power consumption. Federated learning is a continuous process, so that finding a low power device for PPFL is an interesting research topic.

References

Ben-Nun T, Hoefler T (2019) Demystifying parallel and distributed deep learning: an in-depth concurrency analysis. ACM Comput Surv (CSUR) 52(4):1–43

Bonawitz K, Eichner H, Grieskamp W, Huba D, Ingerman A, Ivanov V, Kiddon C, Konečný J, Mazzocchi S, McMahan HBet al. (2019) Towards federated learning at scale: system design. arXiv:1902.01046

Bonawitz K, Ivanov V, Kreuter B, Marcedone A, McMahan HB, Patel S, Ramage D, Segal A, Seth K (2017) "Practical secure aggregation for privacy-preserving machine learning. In: proceedings of the ACM SIGSAC conference on computer and communications security 2017. pp 1175–1191

Briggs C, Fan Z, Andras P (2020) A review of privacy-preserving federated learning for the internet-of-things. arXiv e-prints, pp. arXiv–2004

Chen Y, Luo F, Li T, Xiang T, Liu Z, Li J (2020) A training-integrity privacy-preserving federated learning scheme with trusted execution environment. Inf Sci 522:69–79

Dean J, Corrado G, Monga R, Chen K, Devin M, Mao M, Ranzato M, Senior A, Tucker P, Yang K et al (2012) Large scale distributed deep networks. In: Advances in neural information processing systems. pp 1223–1231

Grama M, Musat M, Muñoz-González L, Passerat-Palmbach J, Rueckert D, Alansary A (2020) Robust aggregation for adaptive privacy preserving federated learning in healthcare. arXiv:2009.08294

Hsu R-H, Wang Y-C, Fan C-I, Sun B, Ban T, Takahashi T, Wu T-W, Kao S-W (2020) A privacy-preserving federated learning system for android malware detection based on edge computing. In: 2020 15th Asia joint conference on information security (AsiaJCIS). IEEE, pp 128–136

Kaissis GA, Makowski MR, Rückert D, Braren RF (2020) Secure, privacy-preserving and federated machine learning in medical imaging. Nat Mach Intell 2(6):305–311

Li M, Andersen DG, Park JW, Smola AJ, Ahmed A, Josifovski V, Long J, Shekita EJ, Su B-Y (2014) Scaling distributed machine learning with the parameter server. In: 11th {USENIX} Symposium on operating systems design and implementation ({OSDI} 14). pp 583–598

Liu Y, James J, Kang J, Niyato D, Zhang S (2020) Privacy-preserving traffic flow prediction: a federated learning approach. IEEE Internet Things J 7(8):7751–7763

Liu X, Li H, Xu G, Lu R, He M (2020) Adaptive privacy-preserving federated learning. Peer-to-Peer Netw Appl 13:2356–2366

Lu X, Liao Y, Lio P, Hui P (2020)Privacy-preserving asynchronous federated learning mechanism for edge network computing. IEEE Access 8:48 970–48 981

Lyu L, Yu J, Nandakumar K, Li Y, Ma X, Jin J, Yu H, Ng KS (2020) Towards fair and privacy-preserving federated deep models. IEEE Trans Parallel Distrib Syst 31(11):2524–2541

McMahan B, Moore E, Ramage D, Hampson S, y Arcas BA (2017) Communication-efficient learning of deep networks from decentralized data. In: Artificial intelligence and statistics. PMLR, pp 1273–1282

Sav S, Pyrgelis A, Troncoso-Pastoriza JR, Froelicher D, Bossuat J-P, Sousa JS, Hubaux J-P (2020) Poseidon: privacy-preserving federated neural network learning. arXiv:2009.00349

Wang Z, Song M, Zhang Z, Song Y, Wang Q, Qi H (2019) Beyond inferring class representatives: User-level privacy leakage from federated learning. In: IEEE INFOCOM 2019-IEEE conference on computer communications. IEEE, pp 2512–2520

Zhang X, Fu A, Wang H, Zhou C, Chen Z (2020) A privacy-preserving and verifiable federated learning scheme. In: ICC 2020-2020 IEEE international conference on communications (ICC)

Chapter 6
Attacks on Deep Learning and Their Countermeasures

Abstract This chapter categorizes the types of the adversarial model on privacy-preserving deep learning based on its behavior, define the major security goals of privacy-preserving deep learning for machine learning as a service, discuss the possible attacks on privacy-preserving deep learning for machine learning as a service, and provide detailed explanations about the protection against the attacks.

6.1 Adversarial Model on PPDL

The attacks in PPDL is keep growing and evolving over time. The adversaries always try to find a new attacking method modified from or different to the previous attacks, while the defenders keep trying to defend against any kind of new attacks that is very difficult to expect in advance. This is a never ending hide-and-seek game. Usually, there are some reasonable assumption on the power of adversaries in order to classify their attack methods to be successful or not. This is called the adversarial model. The adversaries may be assumed to access the attacking targets like the training model by black-box or white-box (or grey-box) model. Black-box model means that the adversaries are only able to do queries to the server and receive prediction results with limited access to the internal operation. They do not have direct access to the training model. While white-box (or grey-box) model indicates that the adversaries can download the training model with all (or limited) the hyper-parameters and others. We categorize the adversarial model in PPDL based on the adversary's behavior, adversary's power, and adversary's corruption types as shown in Fig. 6.1.

6.1.1 Adversarial Model Based on the Behavior

We categorize the adversarial model based on the behavior into either honest-but-curious or malicious.

K. Kim and H. C. Tanuwidjaja, *Privacy-Preserving Deep Learning*,
SpringerBriefs on Cyber Security Systems and Networks,
https://doi.org/10.1007/978-981-16-3764-3_6

Fig. 6.1 Adversarial model
in PPDL

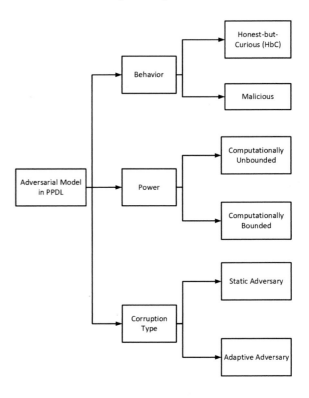

- Honest-but-Curious(HbC): In an Honest-but-Curiousc(HbC) model, assuming that
  all parties, including the corrupted party, follow the security protocol honestly but
  they do not pursue any malicious activity toward the system or other participants.
  However, the corrupted party tries to perform a "curious action" to learn or extract
  sensitive information from the model or from the other participants. This model is
  one of the most popular and reasonable adversaries used in PPDL like other areas
  since it is easy to scrutinize all kind of the attacks.
- Malicious: This scenario is also known as the active adversary model because
  the corrupted parties will actively try to attack even if they must deviate from
  the existing security protocol. If the corrupted parties can prematurely halt their
  attacking process, sometimes the model is also recognized as a fail stop model. But
  this is very strong assumption which is extremely difficult to validate the security
  of the system.

6.1.2 Adversarial Model Based on the Power

We can categorize the types of adversarial model based on their capability with
computationally unbounded or bounded power.

- Computationally Unbounded: This means that the adversary has unlimited computational power. As a result, it is considered as the ideal adversary. It is usually used in theoretical information security field as it does not exist in real life.
- Computationally Bounded: This means that the adversary has limited computational power. Usually, based on the difficulty of the underlying cryptographic problem, the time complexity during the attack process is limited to the polynomial time. If the adversaries have no power to the limit of the bounded complexity, we can claim the a system can be secure from the sense of computational power.

6.1.3   Adversarial Model Based on Corruption Type

We categorize the adversarial model based on the type of the corruption into static adversary and adaptive adversary.

- Static Adversary: In this model, the corrupted parties are defined before the protocol starts. An honest party will always stay honest, and a corrupted party will always stay corrupted.
- Adaptive Adversary: In this model, an adversary will decide which party to corrupt based on the current situation. As a result, an honest party can become corrupted in the middle of protocol execution. However, in the adaptive model, an adversary can change the corrupted party such that the corrupted party can become honest again. This is classified as an adaptive-mobile adversary. This is very strong type of adversary relative to the static one.

6.2   Security Goals of PPDL

PPDL solutions on DL-as-a-service frameworks have three major security goals to avoid the abuses of their privilege happened during the model training.

- **No direct access to training data**: The first goal is to prevent the server from acquiring the training data in the training phase which would be sensitive data owned by the client. All PPDL schemes contain privacy measures to prevent the direct leakage of the training data. HE- and MPC-based approaches solve this issue by encrypting and distributing the training data, respectively. Some methods allow to perform lower-layer calculations in the client side while hardware-based approaches encapsulate lower-layer calculations inside some confidential environment.
- **No access to data during prediction**: The second security goal of PPDL aims to prevent the server from directly acquiring the input to the model in the prediction phase. In most cases, this goal can be achieved together with the first goal. This goal is only applied when the client delegates prediction to the server.

- **Prevent advantage of white-box access**: The third goal is to prevent the server from taking advantage of white-box access of the model. With the white-box access on a model, a server (as an adversary) may deploy several known attacks which are known to be easy on the white-box assumption. As DNNs tend to have more parameters than other machine learning algorithms due to the hidden layers, black-box models could retain more information on training data.

6.3   Attacks on PPDL

6.3.1   Membership Inference Attack

Generally, membership inference means deciding whether given data were used for generating aggregation of the data or not. Membership inference attack is an attack that an adversary tries to infer if a particular data was used to train a deep learning model. In the context of deep learning, a model itself (including the model parameters) can be regarded as the 'aggregation' of the training data. Therefore, membership inference attacks on DL models indicate attacks to decide whether given data belong to the training dataset (or not). Shokri et al. (2017) provided one of the first suggestions of membership inference attacks.

Membership inference attacks is to aim the attacks for the models violating the security goal of PPDL. Stronger versions of membership inference attacks include extraction of some properties of sensitive training data or even recovery of the training data, which can be reduced to normal membership inference attacks. Usually, membership inference attacks harness overfitting during the training, producing a difference in accuracy between the training data and the other data. Some defensive mechanisms dedicated to membership inference have been proposed including dropout (Salem et al. 2018) and adversarial regularization (Nasr et al. 2018).

For membership inference attacks in cryptography-based PPDL models, the adversary can only exploit the black-box access of the training model. Thus, the adversary cannot obtain the training parameters used for the model in a cleartext form. On the other hand, for membership inference attacks against secure enclaves-based PPDL models, the adversary can exploit the white-box access of the training model. As a result, the adversary can access the used parameters of the training model easily.

For DP-based models, the trade-off between the model accuracy and the membership inference attacks according to the selection of the privacy parameter has been studied (Rahman et al. 2018). Appropriate choices of the privacy parameter result in moderate utility with low accuracy for the membership inference attacks. However, further experiments are required for the extensibility of their analysis toward other types of tasks outside image classification.

6.3.2 Model Inversion Attack

As an attack toward the models to avoid the second security goal of PPDL, "**No access to data during prediction**", a model inversion attack can happen during the prediction phase introduced by Fredrikson et al. (2014, 2015). Given the non-sensitive features of the original input data and their prediction results for a model, model inversion attacks aim to find the sensitive features of the input data. It means that if an adversary has a knowledge of model's prediction, the training model can be inverted.

In cryptography-based and secure enclaves-based PPDL models, we expect a similar advantage as that of membership inference attacks. For DP-based models, there has been limited research on the trade-off between the model accuracy and the attack performance, such as the analysis by Wang et al. (2015) against regression models. Although a similar analysis for differentially private DL models remains for future work, we expect a similar trade-off.

6.3.3 Model Extraction Attack

Model extraction attacks (Tramèr et al. 2016), also known as model-stealing attacks, are attacks toward the third security goal of PPDL. When a black-box (target) model is given, the objective of model extraction attacks is to construct a model equivalent to the target model. Once an adversary succeeds with a model extraction attack, the adversary then accesses a white-box model. The attacker can take direct advantage of the model if the model owner sells the model access. The obtained model also becomes a "stepping stone" (Zhao et al. 2019) toward further attacks utilizing white-box models.

Model extraction attack means that an adversary tries to learn an approximation of model by doing many queries to the server. It can also be used for model inversion that an attacker learns the approximation of a model then attempts to reconstruct its training data. Again, adversarial servers against cryptography-based DL models have negligible advantages over those of outsiders, excepting that of the model structure. Without the help of the client, the server cannot obtain the decrypted parameter values of the model.

Most differentially private models and hardware-based PPDL models do not fulfill the third security goal, as they reveal model parameters to the server. The extraction of such PPDL models is meaningless for the adversarial servers, as the servers already have the white-box access on the models, which is the purpose of the attack. Although the servers possess some part of the model parameters, it is relatively easy to extract the remaining parameters by observing the intermediate activation levels.

Table 6.1 PPDL as countermeasures against attacks on DL models

PPDL types	Cryptography-based	DP-based	Secure enclaves-based
Membership inference attack	O	Δ	X
Model inversion attack	O	Δ^*	X
Model extraction attack	O	N/A	N/A

O: An effective countermeasure for the given attack

X: An ineffective countermeasure for the given attack

Δ: The effectiveness of the defense against the given attack has a trade-off with the privacy-preserving parameters of DL models

Δ^*: This trade-off has been confirmed for non-DL models, and it is expected to be the same for DL models

N/A: The adversarial assumption of the given attack is not applicable for the PPDL method

6.4   Countermeasure and Defense Mechanism

Many cryptography-based approaches defined in Chap. 2 achieve the third goal by keeping the parameters encrypted. However, some PPDL models do not assume this third goal and allow the server to access the plaintext parameters of the model. For instance, DP-based models allow white-box access, but the schemes aim to make the information extractable from the parameters negligible.

Although there are many other types of attacks on DL models, in this section we only discuss the attacks that can be mitigated by some of the PPDL approaches. In other words, the following attacks are related to one of the goals of PPDL. Table 6.1 provides a brief summary on PPDL as a countermeasure against attacks.

References

Fredrikson M, Jha S, Ristenpart T (2015) Model inversion attacks that exploit confidence information and basic countermeasures. In: Proceedings of the 22nd ACM SIGSAC conference on computer and communications security. pp 1322–1333

Fredrikson M, Lantz E, Jha S, Lin S, Page D, Ristenpart T (2014) Privacy in pharmacogenetics: an end-to-end case study of personalized warfarin dosing. In: 23rd {USENIX} Security symposium ({USENIX} security 14). pp 17–32

Nasr M, Shokri R, Houmansadr A (2018) Machine learning with membership privacy using adversarial regularization. In: Proceedings of the 2018 ACM SIGSAC conference on computer and communications security. pp 634–646

Rahman MA, Rahman T, Laganière R, Mohammed N, Wang Y (2018) Membership inference attack against differentially private deep learning model. Trans Data Priv 11(1):61–79

Salem A, Zhang Y, Humbert M, Berrang P, Fritz M, Backes M (2018) Ml-leaks: model and data independent membership inference attacks and defenses on machine learning models. arXiv:1806.01246

Shokri R, Stronati M, Song C, Shmatikov V, " (2017) Membership inference attacks against machine learning models. In: IEEE symposium on security and privacy (SP). IEEE, pp 3–18

Tramèr F, Zhang F, Juels A, Reiter MK, Ristenpart T (2016) Stealing machine learning models via prediction apis. In: 25th {USENIX} security symposium ({USENIX} security 16). pp 601–618

Wang Y, Si C, Wu X (2015) Regression model fitting under differential privacy and model inversion attack. In: Twenty-fourth international joint conference on artificial intelligence

Zhao J, Chen Y, Zhang W (2019) Differential privacy preservation in deep learning: challenges, opportunities and solutions. IEEE Access 7: 48 901–48 911

Concluding Remarks and Further Work

A comprehensive survey on recent advances of privacy-preserving deep learning has been conducted by covering the fundamental concepts of classical privacy preserving technologies and the key understanding of deep learning from the latest top-quality publications. The classical privacy-preserving methods can utilize group-based anonymity, cryptographic method, DP, and secure enclaves or its hybrid. The remarkable progress of X-based PPDL that includes HE-based, secure MPC-based, DP-based, secure enclaves-based, and hybrid-based PPDL is summarized from the up-to-date public literature and followed by the pros and cons of each method, providing the comparison, and discussing their weaknesses and countermeasures of each PPDL method. But more *exact measure of evaluation* may be considered.

To avoid the single point of failure in a centralized PPDL, the new direction of PPDL becomes to utilize the distributed or federated learning. The recent developments on PPFL are also surveyed by dividing the function specific and application specific publications. Application specific PPFL tends to advance by adding new features and meeting strict cryptographic requirements that is similar to the progress of the multi-party cryptographic protocol in 1990s. New attacks on deep learning and their countermeasures are suggested as selection criteria for your application.

As further work, getting the efficiency and scalability issues together including privacy-preserving will be one of challenging issues. Implementing the best practices of a specific application PPFL like healthcare, cloud computing, market prediction, banking, e-commerce, *etc.* and finding the optimal aggregation method from partial gradients for PPFL are another challenges. PPFL ensuring the strict fairness even if the intermediate steps are interrupted during PPDL or PPFL processing and providing individual and universal verifiability for all participants are challenging issues too.

The concepts of AI is rapidly advancing and evolving such as *explainable* AI referring to that the results of the solution must be understood by humans. Along this, *explainable* PPDL is also investigated.

K. Kim and H. C. Tanuwidjaja, *Privacy-Preserving Deep Learning*,
SpringerBriefs on Cyber Security Systems and Networks,
https://doi.org/10.1007/978-981-16-3764-3

Finally, we hope that this monograph not only provides a better understanding of PPDL and PPFL but also facilitates future research activities and application development.

Printed in the United States
by Baker & Taylor Publisher Services